Primary Science Learning for Children, Teachers, and Communities

Sarah Earle · Christine Preston · Helen Georgiou ·
Ange Fitzgerald
Editors

Primary Science Learning for Children, Teachers, and Communities

Stories of Practice and Possibility for Science Educators

Springer

Editors
Sarah Earle
Bath Spa University
Bath, UK

Christine Preston
The University of Sydney
Hornsby, NSW, Australia

Helen Georgiou
University of Wollongong
Wollongong, NSW, Australia

Ange Fitzgerald
RMIT University
Melbourne, VIC, Australia

ISBN 978-981-95-0560-9 ISBN 978-981-95-0561-6 (eBook)
https://doi.org/10.1007/978-981-95-0561-6

This Springer imprint is published by the registered company Springer Nature Singapore Pte Ltd.
The registered company address is: 152 Beach Road, #21-01/04 Gateway East, Singapore 189721, Singapore

Contents

Contributors

Linda Ahrenkiel UCL University College, Odense, Denmark

Margien Bakker IPABO University of Applied Sciences, Amsterdam, The Netherlands

Lonneke Boels Hogeschool Utrecht, University of Applied Sciences, Utrecht, The Netherlands

Ondine Bradbury Swinburne University of Technology, Melbourne, Australia

Jared Carpendale Te Kura O Te Mātauranga/Institute of Education, Te Kunenga Ki Pūrehuroa/Massey University, Palmerston North, New Zealand

Katherin Cartwright University of Sydney, Sydney, Australia

Connie Cirkony Faculty of Education, Monash University, Melbourne, VIC, Australia

Simon Crook The University of Sydney, Sydney, Australia;
CrookED Science, Dulwich Hill, Australia

Liliane Dionne University of Ottawa, Ottawa, Canada

Sarah Earle Bath Spa University, Bath, UK

Anika Embrechts Regional Community College of Twente, Hengelo, The Netherlands

Ange Fitzgerald RMIT University, Melbourne, Australia

Vilma Galstaun University of Sydney, Sydney, Australia

Helen Georgiou University of Wollongong, Wollongong, NSW, Australia

Regula Grob University of Teacher Education Lucerne, Lucerne, Switzerland

Anna Hotze Delft University of Technology, Delft, The Netherlands

Helle Hovgaard Jørgensen UCL University College, Odense, Denmark

Vincent Jonker Utrecht University, Utrecht, The Netherlands

Ronald Keijzer IPABO University of Applied Sciences, Amsterdam, The Netherlands

Mijung Kim Faculty of Education, University of Alberta, Edmonton, AB, Canada

Remke Klapwijk Delft University of Technology, Delft, Netherlands

Roslynn Brain McCann Utah State University, Moab, USA

Anna-Vera Meidell Sigsgaard Department of Teacher Education, University College of Copenhagen, Copenhagen, Denmark

Melanie Nash RMIT University, Melbourne, Australia

Thabisile Nkambule Curriculum Studies and Social Science Department, Wits School of Education, University of the Witwatersrand, Johannesburg, South Africa

Morten Rask Petersen UCL University College, Odense, Denmark

Christine Preston Sydney School of Education and Social Work, The University of Sydney, Sydney, NSW, Australia

Doras Sibanda Science and Technology Department, School of Education Edgewood Campus, University of KwaZulu-Natal, Pietermaritzburg, South Africa

Karyn Saunders Victoria University of Wellington, Wellington, New Zealand

Carolyn Julie Swanson School of Education, Auckland University of Technology, Auckland, New Zealand

Kim Chwee Daniel Tan National Institute of Education, Nanyang Technological University, Singapore, Singapore

Brian Tweed Te Kura O Te Mātauranga/Institute of Education, Te Kunenga Ki Pūrehuroa/ Massey University, Palmerston North, New Zealand

Jennifer Way University of Sydney, Sydney, Australia

Kevin J. White University of Toronto, Toronto, Canada

Jennifer Yeo Singapore University of Social Sciences, Singapore, Singapore

Symen van der Zee Saxion University of Applied Science, Enschede, The Netherlands

Caiwei Zhu Delft University of Technology, Delft, Netherlands

Sphamandla Zulu Science and Technology Department, Wits School of Education, University of the Witwatersrand, Johannesburg, South Africa

Chapter 1
Introduction: Contextual Shifts—Reframing Primary Science Education Through Stories of Practice

Ange Fitzgerald, Sarah Earle, Christine Preston, and Helen Georgiou

Abstract This book, *Primary Science Learning for Children, Teachers and Communities: Stories of Practice and Possibility for Science Educators*, offers a transformative exploration of primary science education by documenting diverse narratives of success from educators worldwide. Moving away from the common deficit-focused view of science education, this collection emphasises the learning successes and strengths of children, teachers and communities. By showcasing positive stories, we demonstrate how primary science can foster critical thinking, creativity and interdisciplinary learning. In addition, the book highlights community engagement and professional development as essential elements of meaningful science education. Through illustrative case studies, readers gain practical insights into the challenges and triumphs experienced by primary science educators. This collection provides globally relevant perspectives on how to overcome obstacles, create engaging learning environments, and support children, teachers and their communities in their pursuit of scientific knowledge, skills and understandings. The book ultimately serves as a resource for educators, researchers and policymakers to drive innovation and foster a more inclusive approach to primary science education.

Keywords Primary science · Stories of practice · Strength-based

A. Fitzgerald
RMIT University, Melbourne, Australia

S. Earle (✉)
Bath Spa University, Bath, UK
e-mail: s.earle@bathspa.ac.uk

C. Preston
University of Sydney, Sydney, NSW, Australia

H. Georgiou
University of Wollongong, Wollongong, NSW, Australia

S. Earle et al. (eds.), *Primary Science Learning for Children, Teachers, and Communities*, https://doi.org/10.1007/978-981-95-0561-6_1

1.1 *Introducing*: Why This Work Matters

Through the inception of this collection—*Primary Science Learning for Children, Teachers and Communities: Stories of Practice and Possibility for Science Educators*—we intend to provide an inspiring and constructive framework for the learning and teaching of primary science. By primary science, we mean the science teaching that takes place in primary or elementary school. The exact age varies in each context, but this is broadly from age four or five to age 11 or 12. The exact content of primary science also varies across contexts, with some countries focused on the natural sciences of biology, chemistry and physics, while others take a broader curricular view, exploring integrated Science, Technology, Engineering and Mathematics (STEM) or including areas of humanities like geography. Interdisciplinary links may also be noted, for example, by the inclusion of the Arts in STEAM. Thus within this volume, 'primary science' may be conceived more broadly or narrowly in each chapter, depending on the context within which the chapter is set.

The need for this collection arises from a growing recognition that primary science education is often framed and modelled in deficit terms (Fitzgerald & Smith, 2016; Nag Chowdhuri et al., 2021). As a counterpoint, we offer a compilation of 'success' stories—narratives that celebrate the innovations, strengths and potential within the field of primary science education. Through this book, educators and educational researchers from across the world have come together to share their real-life experiences and evidence-based insights that highlight the transformative power of reframing primary science education with a focus on the inherent possibilities and opportunities. Engaging with children, teachers and communities to support the agency of each party, so that they feel that science is 'for them' (Nag Chowdhuri et al., 2022). The accounts within this collection draw from diverse global contexts, demonstrating that while challenges do exist, they can be met with ingenuity, resourcefulness and optimism by those dedicated to making primary science accessible and engaging for young learners, their teachers and their wider communities.

A distinguishing feature of this book is the emphasis on stories as a pedagogical tool (Xusnida, 2023). By presenting a series of reflective, narrative-driven chapters, we offer insights into the experiences of children, teachers and communities working within and towards improved primary science education. These stories are not only informative but also inspire a shift in perspective from viewing primary science education as a field in crisis, to one rich with opportunities for interdisciplinary learning, creativity and community involvement (Novis-Deutsch et al., 2024).

1.2 *Positioning*: Why This Collection and Why Now

Common widespread and worldwide challenges, such as a lack of teacher confidence, insufficient resourcing and competing curricular priorities and policies, are well-documented (Fitzgerald et al., 2024; Wellcome, 2017). Recent developments, including the pushback against inquiry-based instruction, the rise of artificial intelligence in education and increasing emphasis on international standardized assessments (e.g., PISA, TIMSS), have further compounded these challenges. These dynamics have triggered the listing of issues (Bianchi et al., 2021) and extensive critique, which often reinforces a negative narrative around the state of primary science.

Alongside these challenges, however, there are numerous stories of success connected with primary science education—stories that reflect the creativity, dedication and resilience of educators. This collection seeks to shine a spotlight on these successes, celebrating the strengths and innovations present in contemporary approaches to primary science learning and teaching. By adopting a positive, strengths-based approach to frame this collection, we are contributing to a significant and growing body of literature in education that emphasises the importance of recognising and building upon what has been found to work in particular contexts.

This collection is particularly timely given the significant reforms and shifts taking place in both school-based education and initial teacher preparation globally (Gouëdard et al., 2020). As the focus on STEM education intensifies, primary science education has gained renewed attention. In addition, the rise of global challenges such as climate change, sustainability and technological advancement makes the learning and teaching of science in the early years of education more critical than ever before (Kumar et al., 2023). In response to these developments, this book is positioned to provide a hopeful, futures-oriented perspective. It is a viewpoint that aligns with contemporary educational goals and the need for innovative, student-centred agentic approaches to learning and teaching.

By documenting successful practices, we intend to respond to the growing demand for professional development and learning resources that are evidence-based and reflective of current real-world teaching experiences. As educators grapple with rapidly changing educational landscapes, this collection provides a much-needed resource for those seeking inspiration and practical guidance on how to foster deeper student engagement and understanding of science.

This collection is situated within the broader context of the Anthropocene, a description of our current epoch, in which human activity is having a profound and lasting impact on the Earth and its systems. The numerous global issues we face, such as environmental degradation and biodiversity loss, call for educational approaches that equip children not only with scientific knowledge but with the capacity to make sense of complex issues and take informed, ethical action. Within this framework, primary science education has a vital role to play in nurturing these capabilities from the early years, helping young learners understand their place within interconnected natural and social systems. By recognising the urgency and possibilities of science learning and teaching in the Anthropocene, we strengthen the case for primary science education that is hopeful, responsive and deeply connected to children's lives and futures (Gienger et al., 2024).

1.3 *Contributing*: Why Story Practice

The decision to organise this collection around stories of practice reflects our belief in the power of storytelling as a tool for meaningful sharing and professional learning (Xusnida, 2023). Storytelling allows educators to craft their experiences in a way that resonates with readers, offering both practical insights and emotional connections. In the context of primary science education, where teachers can often experience a sense of isolation in their challenges, sharing stories of practice helps to build a sense of community through shared experience (Fitzgerald & Smith, 2016).

In disciplines such as tourism and business, storytelling has been employed to provide a socially constructed, interpretive dimension to professional practices (Dredge, 2011). Similarly, in education, storytelling allows for a flexible and accessible approach to presenting research and practice, enabling authors to communicate their findings in ways that speak directly to the reader. By drawing on the stories of educators and education researchers working in diverse contexts, this collection enriches the body of knowledge around primary science learning and teaching, which provides a platform for under-represented voices, particularly those from non-metropolitan and marginalised communities.

Sharing stories of practice also allows for the incorporation of various epistemologies, methodologies and analytical frameworks. This diversity of approaches enhances the richness of the collection, as each chapter offers unique perspectives and insights into how science is both learned and taught in different settings. The stories within this book are not only descriptive but also reflective, which has been explicitly designed to encourage readers to critically engage with the stories and consider how it might apply to their own practice.

By focusing on stories of practice in science education, we also emphasise the importance of context in science learning and teaching. Each story is grounded in the specific realities of a particular classroom, school or community, offering a nuanced understanding of the challenges and opportunities that can be present within primary science education. The intention is that teachers, teacher educators and educational researchers will find in these stories a mirroring of their own experiences, as well as new ideas and strategies for enhancing their practice.

1.4 *Impacting*: Why This Collection is Relevant and to Whom

The implications of this collection for science teachers, teacher educators and educational researchers are profound and far-reaching. First and foremost, *Primary Science Learning for Children, Teachers and Communities: Stories of Practice and Possibility for Science Educators* provides educators with a unique opportunity to reflect on their own practice, in light of the experiences and successes shared by their peers. By framing reflection within this positive context, we aim to support teacher agency, for teachers to feel that they can explore the possibilities for their practice. 'Agency is not something that people can have – as a property, capacity or competence – but is something that people do' (Biesta et al., 2015, p. 626) and we want educators to be inspired to 'do'. The stories within this collection demonstrate the potential for transformation in primary science education when innovative approaches and positive perspectives are applied.

For primary science teachers, the practical examples and case studies offered in this book serve as a source of inspiration. The emphasis we place on 'success' stories enables teachers to shift their mindset from one of focusing on limitations and barriers to one of recognising possibilities and opportunities for science education in their classrooms. By reading about the creative solutions employed by their counterparts in various diverse contexts and global communities, educators can explore new strategies that might be applicable to their own science teaching approaches. This, in turn, can lead to more engaging and effective science learning, with a stronger emphasis on inquiry, critical thinking and hands-on learning. Empowered and agentic teachers can become 'change agents' (Buchanan et al., 2020), developing their own practice and sharing across their schools.

The collection also holds significant implications for teacher educators. As they prepare the next generation of teachers, they can use this collection as a resource to foster confidence and innovation in their own students—our future teachers of science. The stories of practice embedded in this collection highlight the importance of professional development and reflective practice, encouraging teachers to continually evolve in their approach to primary science education. The narrative approach used in the book can also be a powerful tool for teacher education programs, as it encourages pre-service and in-service teachers alike to view their own experiences as valuable sources of learning and growth.

By drawing on the strengths-based approach to science education, this collection also challenges the deficit narratives that often dominate discussions in teacher education. It encourages teacher educators to rethink how they frame the challenges of teaching primary science and to focus on the assets and resources that teachers bring to the classroom. Ultimately, this book empowers both teachers and teacher educators to embrace a more positive, hopeful view of primary science education; one that prioritises student engagement, creativity and community involvement.

Beyond the classroom, the implication of this work extends to policymakers and educational leaders in a range of contexts, including community-based settings. The stories of practice powerfully provide evidence that, despite the challenges faced by primary science educators, there are pockets of excellence and innovation that can serve as models for broader systemic change. By promoting a positive, strengths-based and place-oriented approach to primary science education, this collection offers a blueprint for policies and practices that support teacher autonomy, professional development and learning, and interdisciplinary learning. Policymakers will find valuable insights into the kinds of support systems and resources that are necessary to create more vibrant and successful primary science education programs.

1.5 *Structuring*: Why This Book is Organised in the Way It is

The structure of *Primary Science Learning for Children, Teachers and Communities: Stories of Practice and Possibility for Science Educators* is intentionally designed to reflect the holistic and interconnected nature of primary science education. We have organized the book into three main sections, each focusing on a different group of stakeholders: children, teachers and communities. These sections represent the key elements that shape and influence the practice of science education in primary schools, and they offer readers a comprehensive view of how science teaching and learning unfold in diverse contexts.

The Part I focuses on children, placing them at the centre of the discussion. This choice reflects our belief that primary science education should always be rooted in the experiences, needs and interests of young learners. By starting with the child, we emphasize the importance of seeing science education from a student-centred perspective. The chapters in this section explore how children engage with scientific concepts and skills through play, hands-on exploration and multimodal communication. By examining science education from the child's point of view, we aim to highlight the importance of fostering curiosity and a love for learning in the early years of education.

The Part II focuses on teachers, who are the facilitators of quality primary science education. This section delves into the challenges and opportunities faced by teachers as they work to implement effective science instruction. The chapters here emphasise the importance of professional learning, reflective practice and collaboration among educators. By presenting stories from teachers who have navigated the complexities of teaching science, this section provides valuable insights into how teachers build a STEM identity and can be supported to become more confident and effective in their practice.

The Part III of the book turns to the broader context of communities, recognising that primary science education does not happen in isolation and is in fact directly embedded in the notion of place. Communities play a crucial role in shaping how science is taught and learned, particularly in contexts where resources are limited, or cultural and environmental factors come into play. The chapters in this section examine the role of local and global communities in supporting access to science education, with a particular focus on Indigenous knowledge, sustainability and the connections between science and everyday life. By exploring these themes, we aim to show how primary science education can be enriched by drawing on the knowledge and resources of the wider community.

The development of these themes was guided by a desire to present a comprehensive and inclusive view of primary science education. We wanted to create a book that not only addresses the pedagogical and curricular aspects of teaching science but also considers the social, cultural and environmental contexts in which science education takes place. In doing so, we hope to provide a resource that is relevant to educators, researchers and policymakers working in a wide range of settings.

In essence, the themes of children, teachers and communities represent the interconnected nature of primary science education. Each chapter within these sections offers a different perspective on how science is learned, taught and experienced. In collaboration, they form a cohesive narrative that showcases the richness and diversity of primary science education around the world.

1.6 *Organising*: What the Chapters Cover

We begin the collection with consideration of science in early childhood through the Chapter 2 contribution of Morten Rask Petersen, Linda Ahrenkiel and Helle Hovgaard Jørgensen. They explore three approaches enacted by practitioners in their *Science and Play* project, describing *show science, slow science* and *spontaneous science* opportunities. Exploring concrete activities through a lens of Kolb's experiential learning theory (Kolb, 2014) provides the theoretical underpinning for active experimentation, with early childhood practitioners scaffolding reflective observation and dialogue to support early science conceptual development in playful environments.

Following on, Chapter 3 reports on Jennifer Yeo and Kim Chwee Daniel Tan's development of a formative assessment sequence intended to develop primary students' learning of scientific concepts and representational skills. Their chapter showcases the possibilities in conceptual and representational learning of primary-aged students, drawing on multimodality theory and presenting a model for how to develop representational awareness. The activities designed for the research in this chapter are creative and versatile tools to encourage both conceptual understanding and develop communication skills.

Remke Klapwijk and Caiwei Zhu's work in Chapter 4 takes spatial reasoning as its focus, considering both its definition and importance in science and design learning. Manipulating practical materials and mental representations spatially is exemplified through discussion of models, diagrams, language and gestures. Examples of spatial thinking in science and design lessons include story-based problem solving and biomedical design challenges. Spatial reasoning is promoted as an under-developed area of research, with much potential to support children in their science and design activities.

In Chapter 5, Anna-Vera Meidell Sigsgaard investigates teaching strategies to support children's understanding of science concepts through a focus on science terminology. The order in which practical activities and reading tasks are scheduled for children impacts both their scientific and language literacy. The concept of semantic gravity from Legitimate Code Theory (Maton, 2014) is applied to case studies of teaching in upper primary science classes on the topic of buoyancy. Teachers gained insight into links between hands-on tasks, language use and students' conceptual understanding of the science content.

Through Chapter 6, the final chapter with a focus on children's learning, Sarah Earle and Regula Grob explore the use of formative assessment in primary science education by highlighting its cyclical nature and the challenges that can disrupt its effectiveness. By examining stories of practice from Switzerland and the United Kingdom, the chapter illustrates how formative assessment can significantly enhance children's learning when implemented successfully. The chapter contributes to understanding of how targeted feedback and responsive teaching can bridge learning gaps, thereby improving both learner outcomes and teacher confidence in primary science education.

In shifting to teacher learning, Melanie Nash, Ondine Bradbury and Ange Fitzgerald use Chapter 7 to present two perspectives of an immersive teacher education course involving postgraduate-level students who bring rich experiences from diverse fields. Aimed at developing expertise as future primary STEM teachers, narrative vignettes feature the voices of the pre-service teachers, and the educators who designed and facilitated the course. Data analysed using the practice architectures framework gave insights into how the course and interactions within the learning environment influenced participant outcomes.

Ronald Keijzer, Anika Embrechts, Anna Hotze, Vincent Jonker, Margien Bakker, Lonneke Boels and Symen van der Zee continue the teacher learning focus in Chapter 8 by exploring STEM education in the context of The Netherlands. Challenges and opportunities include the relationship between mathematics and the other disciplines, together with use of maker education and assessment. The chapter reports on the SPRONG-STEM project, where teachers worked together in Professional Learning Networks (PLNs) to address challenges at a local level. The authors present recommendations for STEM connections in national guidance and the use of PLNs.

The work by Jennifer Way, Katherin Cartwright, Christine Preston and Vilma Galstaun shared in Chapter 9 reports on a significant professional development program for primary STEM teachers in the Australian state of New South Wales. The program is likely to be one of the largest of its kind, involving hundreds of primary school teachers and impacting potentially thousands of students. The program involves collaboration between university researchers in science and education and classroom teachers. The data includes results from a survey, analysis of project presentations and reflections from the program directors, demonstrating the positive impact on participating teachers and students.

Chapter 10 makes a direct contribution to teacher learning by exploring effective professional development (PD) for primary science teachers. This collaboration from Simon Crook and Helen Georgiou emphasises content-focused, collaborative and inquiry-based approaches. This work addresses challenges like adapting PD to different educational contexts and boosting teacher confidence. Through case studies, the chapter argues that adaptive, ongoing support is crucial for enhancing science education, helping teachers to navigate curriculum changes and foster student engagement through meaningful, context-specific strategies.

Moving into the final theme, Chapter 11 begins our section focused on community learning. Sphamandla Zulu, Thabisile Nkambule, Doras Sibanda and Ange Fitzgerald consider the challenges of teaching science in multigrade classes in rural South African schools. Utilising data from lesson observations and video stimulated recall interviews, they explore teacher perspectives on the learning environment and the use of collaborative peer learning. The unique context is exemplified, with multigrade classes emerging both because of low numbers of children in a rural location, and because of low numbers of teachers leading to large classes. Teacher voices are foregrounded, as they unpick the complexities of teaching primary science to multiple grades in the same class.

Chapter 12 furthers understanding of community learning by exploring the integration of permaculture in primary education as a means to decolonize teaching and promote social and environmental responsibility. Roslynn Brain McCann, Kevin White and Liliane Dionne argue for a curriculum rooted in Indigenous wisdom and permaculture design, which highlights the need for transformative education that intertwines Indigenous and Western knowledge. By addressing the challenges in teaching STEM and environmental education, the chapter offers a framework for educators to inspire a generation equipped to tackle environmental and social challenges.

Adding to understandings of community learning, Chapter 13 explores how Connie Cirkony, Mijung Kim, Brian Tweed, Jared Carpendale and Christine Preston have adopted a strengths-based approach to teaching Indigenous and Western Sciences in initial primary teacher education. In forming a community of practice across three countries—Australia, New Zealand and Canada—the story tells how they support each other through reflecting on their journey towards an integrated approach to teaching science in their respective initial teacher education programs.

Finally, in rounding out this collection, Carolyn Swanson and Karyn Saunders examine the challenges faced by lone academics in primary science education within initial teacher education in Aotearoa (New Zealand). Their work emphasises the geographic and professional isolation that teacher educators in this nation face. Their chapter explores how an online Community of Practice (CoP) supported early career academics (ECA) in both their research and teaching practice. Particular attention is given to the importance of belonging to a community, creating knowledge through shared experiences, and connecting with the wider science education community.

1.7 *Synthesising*: Key Messages

Through the three interwoven components making up *Primary Science Learning for Children, Teachers and Communities: Stories of Practice and Possibility for Science Educators*, it is important to reflect on the key message that ties the sections and chapters together: primary science education is a field rich with potential and opportunity. While the challenges faced by educators are real and often substantial, the stories shared within this collection offer a powerful counter-narrative. These stories demonstrate that success is possible when children, teachers and communities work together in creative, resourceful and collaborative ways.

The central theme running through this book is the idea that primary science education can and should be an engaging, meaningful experience for both children and teachers. By adopting a strengths-based agentic approach, we shift the focus away from what is lacking and towards what can be achieved. This shift in perspective is not only empowering for educators but also essential for the future of science education globally. It is through this positive, forward-looking lens that we hope readers will be inspired to rethink and reimagine their own practice. The diverse range of stories presented in this book—from early childhood education to professional learning and community engagement—offers a comprehensive view of the possibilities within primary science education. These stories provide practical insights and strategies, but they also invite readers to reflect on their own experiences, to question their assumptions, and to explore new ways of thinking about science learning and teaching in primary schooling contexts.

Ultimately, the story of primary science education is one of possibility. It is a story of educators who, despite the challenges, are finding innovative ways to make science accessible and relevant to their students. It is a story of communities coming together to support learning and foster a sense of curiosity and wonder in young learners. It is a story of teachers, who, through reflective practice and professional development, continue to grow and evolve in their craft. As we move forward, it is our hope that this book will serve as a resource and inspiration for educators, researchers, community members and policymakers alike. By sharing these stories of practice, we aim to contribute to a global conversation about the future of primary science education—one that is rooted in optimism, innovation and a commitment to creating meaningful learning experiences for all children.

1.8 *Concluding*: What to Take Away

In conclusion, *Primary Science Learning for Children, Teachers and Communities: Stories of Practice and Possibility for Science Educator* brings together a diverse array of perspectives and experiences to create a comprehensive and inspiring resource for educators and researchers. By focusing on positive and uplifting narratives, the book challenges the deficit perspective that so often dominates discussions of primary science education and instead highlights the creativity, innovation and resilience of primary science educators. The stories within this collection demonstrate that, despite the challenges, there are many successes to celebrate in the field of primary science education. From the innovative approaches of early childhood practitioners to the professional growth of teachers and the engagement of communities, this book offers a vision of what is possible when primary science education is approached with optimism and a commitment to fostering curiosity and critical thinking.

By drawing on the strengths-based approach, the book provides readers with practical strategies and insights that can be adapted to a wide range of educational contexts. More importantly, it invites readers to reflect on their own practices and consider how they can contribute to the global effort to improve science education for all students, particularly those in the primary years of schooling. As educators, researchers, community members and policymakers continue to navigate the evolving landscape of primary science education, *Primary Science Learning for Children, Teachers and Communities: Stories of Practice and Possibility for Science Educator* serves as a valuable resource for those seeking to create more engaging, inclusive and impactful learning experiences. This collection is not just about teaching science; it is about inspiring a generation of learners to think critically, act creatively, and engage with the world around them in meaningful ways.

References

Bianchi, L., Whittaker, C., & Poole, A. (2021). *The 10 key issues with children's learning in primary science in England.* The University of Manchester and The Ogden Trust.

Biesta, G., Priestley, M., & Robinson, S. (2015). The role of beliefs in teacher agency. *Teachers and Teaching, 21*(6), 624–640. https://doi.org/10.1080/13540602.2015.1044325

Buchanan, R., Mills, T., & Mooney, E. (2020). Working across time and space: Developing a framework for teacher leadership throughout a teaching career. *Professional Development in Education, 46*(4), 580–592. https://doi.org/10.1080/19415257.2020.1787204

Dredge, D. (2011). Tourism planning and policy: Historical development and contemporary challenges. In D. Dredge & J. Jenkins (Eds.), *Stories of practice; tourism planning and policy* (pp. x–x). Routledge.

Fitzgerald, A., & Smith, K. (2016). Science that matters: Exploring science learning and teaching in primary schools. *Australian Journal of Teacher Education, 41*(4), 64–78.

Fitzgerald, A., Pressick-Kilborn, K., Mills, R., Pfeiffer, L., & Deehan, J. (2024). Contemporary Australian primary science teacher education—Triumphs, tensions and future directions. *Springer*. https://doi.org/10.1007/978-981-97-5660-5

Gienger, A., Nursey-Bray, M., Rodger, D., Szorenyi, A., Weinstein, P., Hanson-Easey, S., & Yoneyama, S. (2024). Responsible environmental education in the Anthropocene: Understanding and responding to young people's experiences of nature disconnection, eco-anxiety and ontological insecurity. *Environmental Education Research, 30*(9), 1619–1649. https://doi.org/10.1080/13504622.2024.2367022

Gouedard, P., Pont, B., Hyttinen, S., & Huang, P. (2020). *Curriculum reform: A literature review to support effective implementation* (OECD Working Paper No. 239). https://one.oecd.org/document/EDU/WKP(2020)27/En/pdf

Kolb, D. A. (2014). *Experiential learning: Experience as the source of learning and development.* FT Press.

Kumar, P., Sahani, J., Rawat, N., Debele, S., Tiwari, A., Mendes Emygdio, A.P., Abhijith, K.V., Kukadia, V., Holmes, K., & Pfautsch, S. (2023). Using empirical science education in schools to improve climate change literacy. *Renewable and Sustainable Energy Reviews, 178.* https://doi.org/10.1016/j.rser.2023.113232.

Maton, K. (2014) *Knowledge and knowers: Towards a realist sociology of education.* Routledge.

Nag Chowdhuri, M., King, H., & Archer, L. (2021). *The Primary science capital teaching approach: Teacher handbook.* University College London.

Nag Chowdhuri, M., King, H., & Archer, L. (2022). The Primary science capital teaching approach: Building science engagement for social justice. *Journal of Emergent Science, 23*, 34–38. https://www.ase.org.uk/resources/journal-of-emergent-science/issue-23/research-review-primary-science-capital-teaching

Novis-Deutsch, N., Cohen, E., Alexander, H., Rahamian, L., Gavish, U., Glick, O., & Mann, A. (2024). Interdisciplinary learning in the humanities: Knowledge building and identity work. *Journal of the Learning Sciences, 33*(2), 284–322. https://doi.org/10.1080/10508406.2024.2346915

National Center for Education Statistics. (2022). *Trends in International Mathematics and Science Study (TIMSS)*. https://nces.ed.gov/timss/

Programme for International Student Assessment (PISA). https://www.oecd.org/en/about/programmes/pisa.html

Wellcome. (2017). *State of the Nation report of UK Primary Science Education*. The Wellcome Trust.

Xusnida, B. (2023). Stories as an effective pedagogical tool. *Pedagog, 6*(6).

Part I
Primary Science Learning for Children

Chapter 2
The S'ence of Early Childhood Science: Learning Science from Different Approaches

Morten Rask Petersen, Linda Ahrenkiel, and Helle Hovgaard Jørgensen

Abstract The chapter will outline the concepts of (1) *show science*, as short term planned science activities, (2) *slow science*, as continuous long-term science activities with reengagement and (3) *spontaneous science*, as unplanned activities caught in the moment. These concepts—all starting with S—is what we call the S'ence [essence] of early childhood science. The concepts were identified in a design-based research project *Science & Play*. Each of the three approaches to science has a different potential for learning science. The chapter will argue on the importance of addressing all three approaches in practice in order to have children both learning and caring about and playing with science. The chapter will outline these together with the potential of combining the three approaches. Each of the approaches will be exemplified with key episodes from the Danish early childhood education and care (ECEC) settings. A central focus, but still different among the three approaches, is the interplay between phenomena and concepts of science. This will also be outlined both theoretically within a framework of Kolb's learning cycle (Kolb, 2014) and *play moods* (Skovbjerg, 2021).

Keywords Early science · Play · Slow science · Show science · Spontaneous science

2.1 Introduction

In play, children curiously investigate their surroundings—and the potential of learning science is always present. Learning science and play are central parts of the Danish Early Childhood Education and Care (ECEC) curriculum for children aged 1–6 years, and it becomes relevant to speak of a child centred playful 'science approach' since 98% of all children aged 3 or more attend ECEC in Denmark. The

M. R. Petersen (✉) · L. Ahrenkiel · H. H. Jørgensen
UCL University College, Odense, Denmark
e-mail: mrpe@ucl.dk

S. Earle et al. (eds.), *Primary Science Learning for Children, Teachers, and Communities*, https://doi.org/10.1007/978-981-95-0561-6_2

question is therefore how do we understand 'a playful science approach' in an early childhood context?

The Danish ECEC often has an approach of addressing a specific phenomenon or doing spectacular experiments, and often science is done in a perspective of adjusting compulsory school science to an early childhood setting (Hansson et al., 2021). Larimore (2020) argues that a more holistic approach to science is needed in ECEC settings. We argue that science in ECEC needs to combine nature and natural phenomena with children's play to ensure their physical and sensory participation.

This chapter focuses on different approaches to science in the context of less formalised learning environments of ECEC settings in Denmark based on play. Children naturally enjoy observing and thinking about nature and natural phenomena (Eshach & Fried, 2005). The ECEC staff play an important part in scaffolding children's interest in nature and natural phenomena. In Denmark ECEC learning environments are based on a national curriculum where play is a part of the common pedagogical foundation for the six fundamental areas, of which science is one (Ministry of Children and Education, 2020). The science component emphasises an investigative approach, which opens for active involvement of nature and natural phenomena in children's play. Play is vaguely defined in the curriculum but is closely related to another crucial part of the foundation – the active involvement of children's perspectives. In this context, play is understood as children's primary practice form, with active participation as a basic condition.

The chapter will outline three approaches to playful science: (1) *show science*, as short-term planned science situations, (2) *slow science*, as continuous long-term science situations with reengagement and (3) *spontaneous science* (the three S's), as unplanned situations caught in play. Each of these approaches to science has a different potential for learning science through play.

2.2 How Children Playfully Learn Science

According to Larimore (2020), early childhood education is traditionally based on a cognitive (Piagetian) and socio-constructivist (Vygotskian) approach. While still inspired by the Piagetian and Vygotskian approach, in this chapter we twist the focus towards a Deweyan and Kolbian approach to learning science because we want to strengthen a curious, tactile and experimental approach to learning science in a playful way.

Dewey (1913) argues that children can have divided or full attention while learning. Divided attention means that the child puts in the exact level of effort needed to solve the problem, while at the same time addressing more stimulating thoughts. This is addressing a central issue in early childhood learning as well as general learning, namely *why* we are doing this particular activity. For whom is this important? If the children only participate with what is necessary, it means that the child is not motivated for the content but motivated for pleasing the ECEC staff. What is needed is to have learning environments in which children engage fully and thereby enjoy both cognitive and affective experiences. In establishing such learning environments play is fundamental. Children need to engage while playing with the content and have tactile experiences with this; such experiences are called educative (Dewey, 1913).

To provide educative playful science learning environments, it is not enough to just support the children with materials. The ECEC staff have a crucial role in understanding and interpretating these experiences and connect phenomena and concepts while playing.

We find such an approach in Kolb's experiential learning theory (Kolb, 2014). Kolb argues that learning occurs in four different phases within an iterative cycle through 1) active experimentation the children experience and 2) concrete experience. These experiences can lead to 3) reflective observation and emerging hypothesis on the phenomena examined. Such reflective observations can then again lead to 4) an abstract conceptualisation. Such concepts open up for new experimentation and another iteration begins. However, we find that learning science in an early childhood setting is messier, where the phases are more intertwined (see Fig. 2.1). This can be because science often is approached in a cognitive, hypothetical understanding (if, then, because) while playing with science is focusing on an experimental phenomenological approach (what if?).

Children are well capable of doing active experimentation and having concrete experiences while playing, but to make such experiences educative requires abstract concepts that they are not aware of and are not capable of constructing by themselves. This necessitates the active engagement/involvement of the ECEC staff both for the

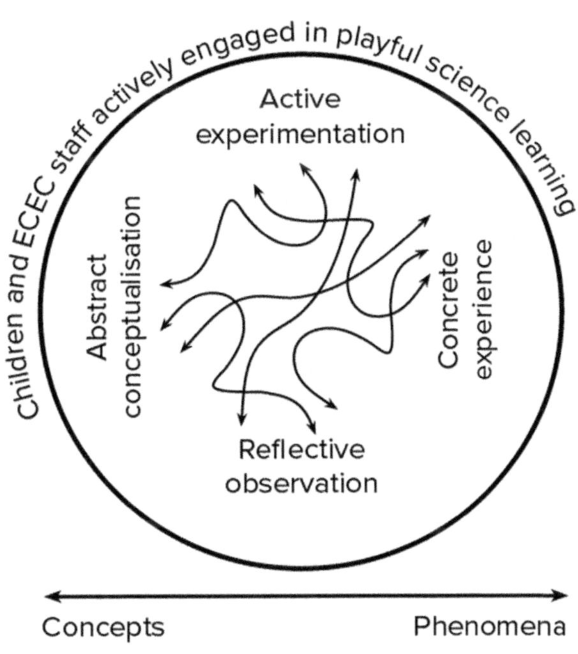

Fig. 2.1 A model of the messy presence and interplay between concepts and phenomena inspired by Kolb

reflective observation to know which questions to ask regarding the experiences and to use scientific concepts to describe the phenomenon investigated (the relation between phenomenon and concepts are outlined below).

In other words, children are capable of having concrete experiences by themselves, but they are not capable of turning those experiences into learning science.

In Fig. 2.1, this is shown as a common engagement of both children and ECEC staff in coupling concepts via reflective observation and abstract conceptualisation with the phenomenon via active experimentation and concrete experiences in relation to play.

2.2.1 Concepts and Phenomena

Colloquially, a phenomenon is how an object appears, how it apparently is and how it appears to us. An object appears in many different ways, in strong or weak lighting, as perceived, imagined, remembered, questioned or communicated. The object might be more or less direct and more or less present (Zahavi, 2018). We can talk about something we have never seen and imagine what it is by hearing the word (e.g. solar eclipse) or a story (who took a bite of the Sun?), seeing a picture or a model. We can also experience a real solar eclipse. When children go through experiences, it is related to phenomena. To convert such experiences into science, we need to conceptualise what is happening in the phenomenon experienced. Following Eshach and Fried (2005, p. 320): '...the way science ultimately allows us to see the world is by providing us with concepts with which we can frame its phenomena'. This means that science is present in the intertwining of phenomena and concepts.

Children experience various phenomena, and such experiences require a language to talk about the phenomena. This could happen either by the children making up their own conceptualisation to describe and explain the phenomena or by the ECEC staff providing the scientific concepts to the children. Vygotsky (1986) describes these two approaches as spontaneous concepts versus scientific concepts. In this chapter, we argue, with offset in Kolb (2014), that scientific concepts combined with concrete experiences is a profitable approach to science in an ECEC setting.

Fleer (2009) describes spontaneous concepts as pyramidic in that they arise from many experiences that are categorised within the same or a related concept like when children are finding beetles, ants and flies and the adults introduce the concept of insects. Opposite to this, scientific concepts can be presented as a downward pyramid, where the general concept covers many phenomena but can only be concretised in one context at a time. This could be when the adult e.g. introduces the concept of friction while children is going down a slide and later introduces friction in another context.

In play, children often spontaneously name scientific phenomena. In the situation below, a group of children are playing at a pond and discover frog eggs; one child spontaneously calls it 'slime', and, in the process of investigating the metamorphosis, they speak of various kinds of transformations.

Relating this to Kolb, there is a need both for conceptualising active experimentation and concrete experiences and for concretising reflective observation and abstract conceptualisation found in scientific concepts.

A playful approach to science involves potentially active experimentation and concrete experience. The abstract conceptualization and reflective observations might be a consequence of professional guidance as we shall see.

2.2.2 Science as the Resource in Play

Play is children's primary practice form and the way they experience the world. Play is by nature dialogic and open-ended (Jørgensen & Skovbjerg, 2024) and when in play you are attuned to each other by saying yes to ideas, moves, materials and people. Following Bogost (2016), play is deliberately working with the material you encounter and taking the dialogic nature of play into consideration all kinds of material (language, senses, moves, space, and 'things') 'whirl' in and out of play in messy ways that support investigative, imaginative and curious processes. Science might as well be relevant (raw) material in play as anything else. Hence, we need to understand better how to frame play in the context with science. Following Skovbjerg (2021, p. 10), 'one is particularly open to new possibilities when playing (…) When one is in a special mood that occurs during play, one can say, "yes" to a lot of things. Play moods are the special frames of mind children has while playing, and they are made by 'both the materials and the actions and forms of exercise during the play situation (Skovbjerg, 2021, p. 11). When having different *play materials* and engaging in different *play practices, play moods* arise. Skovbjerg (2021) presents four different basic types of play practices with play actions, pointing to four basic play moods. Play practices are as follows:

(1) Sliding, where the aim is making as little change as possible from action to action. The play mood is experienced as *devotion,* includes flow and continuity, and the actions are e.g. collecting things to fiddle with or to build or construct something.
(2) Shifting, which can be seen as a strong form of repetition as, for example, in jumping, but the acts of play change through changes in height, speed and direction. The quality of shifting is the unpredictable change, using the body as a tool for the change and opening up for other experiences. The play mood is described as *intensity.*
(3) Displaying, where actions are often related to performing or putting oneself, one's skills or both on stage. The quality of play is related to the expectation of change and the *play mood* is described as experiencing *tension.*
(4) Exceeding, where acts are fleeting and the players expect the play to exceed and change over time, again and again. It is anticipated that the playacting will be 'out of beat' and not about finding a rhythm. Shouting out in enthusiasm,

stories of frivolity are characteristics of play acts that exceed. The experience is described as a *euphoric* play mood.

By including play in science learning processes, we invite the playful 'what-if' in combination with scientific experimentation 'let's find out what happens', and thereby make science the material we deliberately work with and support a curious investigation of science.

2.3 The Method and Context of the Three S's

Data for this chapter are drawn from the research and development project titled *Science and Play* (2020–2023). Three researchers and 15 ECEC centres from three municipalities participated in the project. The project had a design-based research approach (Akker et al., 2006) where ECEC staff were involved in designing experiments as part of professional development (PD), which occurred as a mixture of workshops and field work at the ECEC centres. Both participatory methods were used to collect data (fieldnotes and photo notes); (Emerson et al., 2011; Gulløv & Skreland, 2016). Data collected in *Science and Play* resulted in key episodes wherein children and ECEC staff were engaged in different types of science situations, which is used in this chapter. All data collected was in accordance with the principles in the UNICEF Responsible Data for Children initiative (RD4C).

As part of the *Science and Play* project, a survey was developed to collect qualitative data of the ECEC staff from three participating municipalities (n = 149). The survey is about how they conceive science and play in ECEC settings, with the focus being is on the question, 'When occupied with science, in what way do you do it?' A thematic analysis (Braun & Clarke, 2006) was adopted to analyse the responses collected from the open-ended questions.

2.4 Findings

The thematic analysis of the data revealed the respondents' ways of being occupied with science in ECEC settings. Two themes emerged: planned science and spontaneous science. The theme of planned science was further bifurcated in two significantly different themes: *show science* and *slow science*. The themes that emerged based on the thematic analysis as key episodes were as follows:

1. *Show science* is a short-term planned situation with play qualities initiated by the ECEC staff. The responses had an approach of addressing a specific phenomenon or doing spectacular experiments in an ECEC setting, which was also found by Hansson et al. (2021). Examples of *show science* found in the data included *'cola and mentos'* or *'we made volcanoes'*.

2. *Slow science* is characterised by continuous long-term science activities with reengagement, often planned by the ECEC staff. Observations in slow science activities can often be an initiator for play. Examples from the data included '*sprouting of* tomatoes' and '*development of butterflies.*
3. *Spontaneous science* is unplanned activities taking place while playing. Examples in the data are given when the respondents describe how science is initiated by the children as '*something that spontaneously arises because we are inspired by what the children do or ask about*' or as '*something that occurs spontaneously as part of a play'*. *Spontaneous science* combines nature and natural phenomena with children's play to elicit physical and sensory engagement.

In the following we elaborate on each of these key episodes, wherein children and ECEC staff were engaged in different types of science situations. We relate the key episodes to Kolb and play.

2.4.1 Show Science: Solar Eclipse

Show science is primarily driven by the ECEC staff. It involves planning and takes offset in the idea of showing a phenomenon or explaining a concept (Fleer, 2022). From data collected in *Science and Play*, key episodes were identified as moments when children were engaged in *show science* situations.

Lily from an ECEC centre and a researcher decided to engage the children in the natural phenomenon of a solar eclipse. The phenomenon, where the Moon comes between the Earth and the Sun and the Moon casts its shadow on the Earth, was coming up and there was a chance for seeing it in the sky when the children would be at the ECEC centre. Before the day of the of the solar eclipse, Lily read aloud the children's picture book *Who Eats the Sun?* (Sonneborn, 1982) during lunch. On the day of the eclipse, children and the ECEC staff go to a darkened room. Lily has prepared an environment with a miniature solar system by using a big inflatable globus, a theatre lamp (the Sun) and a small ball as the Moon to focus on light and shadows. She starts by showing how the Moon covers the Sun with the help of a globus, a theatre lamp and small balls. She also briefly talks about the Moon phases, which are observable when moving the ball around the globe. Afterwards, flashlights and balls are given to the children to play with and investigate light and shadow.

Lily tells the children that the eclipse is about to begin: 'It is time to go outside', she says. The children rush outside to the playground where the sky is greyish. They are focused on the sky and soon more and more blue sky appears. Lily hands out a set of solar eclipse glasses to everyone, and all eyes are directed towards the Sun. The Sun is circular. 'Oh, is a bit missing?', asks Lily. 'We can't see it', one child shouts, 'There are too many clouds'. For a few minutes, the Sun is not visible, but then it breaks through: 'WOW!', 'Look!', exclaim many of the children while pointing at the Sun. We observe that a large piece of the Sun is missing. It is no longer circular. 'Wow, someone took a bite of the Sun—as in the book', Lily says with enthusiasm. (Fieldnote).

A solar eclipse is spectacular and has the potential to be a remarkable show. The phenomenon appears in a strong light right before the spectators' eyes. For this science situation, the show is planned by Lily, and she provides the children with the right equipment to actively engage in the situation. However, there is not much for children to do in the concrete experience of a solar eclipse. They look and point at the Sun as spectators of a natural phenomenon for a short while. But Lily expands the 'event' to last longer by letting the children perceive the phenomenon in different ways. First, they become used to the rather strange expression 'solar eclipse'. Second, they listen to a story about someone taking a bite of the Sun and thereby add a metaphorical level to the concrete phenomenon. Third, they play out the concepts (light and shadow and the phases of the Moon) with the help of certain materials in a darkened room. Finally, they go through concrete experiences with the real phenomenon. The expansion adds an element of *slow science* to the situation described. Lily deliberately works with different materials for the children to enter the mood of play. She reads a book aloud to involve an imaginative level; she sets up a play environment with a model of the solar eclipse and provides the children with flashlights and balls to play and investigate the scientific concepts of shadow, light and the Moon phases. This invites the children to run around with the flashlight eager to investigate light and shadow. The shifting action is performed by the playful acts of running and jumping while at the same time deliberately working with the materials in hand. This increases an *excited* play mood. When outside, watching the solar eclipse takes the children to a *tense* play mood. The children display their observations by shouting 'Wow!' and 'Look!' and by being spectators of an event that causes their enthusiasm. Their surprise and enthusiasm when the clouds disappear, and the solar eclipse is visible creates a moment where the mind is set for a *euphoric* play mood.

Overall, this relates to Kolb by both the actions of the children and the ECEC staff. The children are actively playing and investigating with the concepts light and shadow, and their imagination about a solar eclipse is activated on different levels. This turns their attention towards the upcoming solar eclipse and prepares them for the concrete experience of a solar eclipse. During the experimentation, Lily has directed their attention to the significance of distance from the ball to the corresponding spot of light to engage reflective observations. The abstract concepts of light, darkness and distance are presented in different ways by Lily. The playful approach to the situation involves both materials, body and senses, and the outbursts of 'Wow!' primarily signify a serious interest in the phenomenon and the potential of learning more about the phenomenon and the concepts in combination.

2.4.2 Slow Science: From Egg to Frog

Slow science is mostly initiated by the ECEC staff for children to engage with natural phenomena and concepts related to them repeatedly. It develops over time (from hours to years), and that might cause curiosity among children and invite them to ask questions about the phenomena (Ahrenkiel et al., 2023a). Hence, *slow science* allows children to engage and play with various phenomena wherein the ECEC staff can provide them with the relevant concepts to frame the phenomena (Eshach & Fried, 2005).

From the data collected in *Science and Play*, key episodes were identified as moments where children were engaged in *slow science* situations. In one such situation, Bjarke (a member of the ECEC staff) and a group of eight children are on a tour to a small pond:

> *It is early spring, and a group of children from an ECEC-centre are heading towards a small pond. They are bringing nets, buckets and plates. At the pond, the children catch many different things. Suddenly Erica shouts out, 'I have caught a lot of slime'. Bjarke and a couple of children come over. 'Wow! Exciting. What do you think it is?' Bjarke asks. The children are silent. 'Let us try to touch it', Bjarke continues. 'Urk. It is slimy', Erika says. 'It feels soft', says Magnus, while Alberte thinks it is similar to glue. Villads is watching—he does not want to touch it.*
>
> *Bjarke finds a magnifying glass. 'Should we look closer at what Erika found?' Some of the slime is put in a small can and exposed on to the tablet's magnifying glass. They talk about what they see. 'It looks like a bunch of eyes', says Alberte. Back at the ECEC centre the slime is put into an aquarium. (Fieldnote).*

For the next few days, the children observe what happens with the 'slime'. Barke reports that the children are preoccupied with the fact that the 'eyes' in the eggs are getting bigger as time passes. After a couple of weeks, the aquarium is full of tadpole. They do not at all look like slime. When the researcher is visiting the daycare again, the following is observed:

> *Many children come by the aquarium to look at the tadpoles swimming around. 'Wow!', the excitement is big that morning. 'Look!', a child points at a tadpole. Bjarke is near the aquarium and uses the concept of tadpole in every sentence where it is meaningful. Another ECEC staff member enters the room and looks in the direction of the aquarium, 'Wow! What is going on here?' A child answers, 'The eggs are transformed into tadpoles'. (Fieldnote)*

Before the summer break, something new happens. The tadpoles grow legs. They are still tadpoles, but suddenly they have two legs. After the summer break, all the tadpoles have four legs and soon they drop their tail. The eggs have transformed into little frogs. The children go to the pond again and return the frogs to their habitat. Next spring the children visit the pond again, and Bjarke reports that he asked the children if they remembered the frogs from last year and ask the children if they believe it is 'their' frogs croaking. He also explains that they are croaking to find each other and make new tadpoles.

Slow science is characterised by continuous, long-term science activities with reengagement. The metamorphosis of a frog follows the four stages of development described above. During a longer period, the children and the ECEC staff follow the development; as one child says, 'the transformation' from little eggs to tadpoles, to froglets to frogs was happening. The trip to the pond is initiated by the ECEC staff, which is characteristic for s*low science.* The trip holds the potential for children to engage with and gain concrete experiences with different phenomena.

According to Kolb (2014), this gives children the opportunities of active experimentation by holding a 'hands-in-the-pond' approach. The children sense the water and the 'slime' in the pond and are given the opportunity to investigate what they find through both words and actions. Bjarke does not explain what Erica has found but invites the children to be curious and investigative by asking open-ended questions. The concrete experiences and the children's engagement initiate a *slow science* situation, where the children meet several phenomena over time: eggs, tadpoles and frogs. The ECEC staff can relate the concepts of metamorphosis in the phenomenon. This supports what Kolb (2014) calls an abstract conceptualisation of, for example, metamorphosis. In *slow science*, this can be supported by the use of photos. The concrete experiences with eggs at the pond might lead to reflective observations and emerging hypothesis on the phenomena examined. In this key episode, the focus was on the actions the children performed at the pond, which relates to active experimentation and concrete experiences (Kolb, 2014). The *slow science* situation holds the potential to support reflective observations and abstract conceptualisation (Kolb, 2014) when the ECEC staff are engaged in the situation, but this took place in continuation of the fieldnotes collected, and we can therefore only indicate the potential of the situation from the perspective of Kolb's experiential learning theory (Kolb, 2014).

The *play mood* in this *slow science* situation appears to be characterised by devotion, and it supports their ongoing investigation both at the pond and subsequently at the aquarium. The ground play action is sliding when the children use nets, buckets and plates at the pond. For a moment, the *play mood* shifts from devotion to tension and even euphoria when Erica shouts out: '*I have caught a lot of slime*' and the group is excited about her findings and burst out 'W*ow!*' and '*Urk. It is slimy*'. During the time when the children follow the development of frogs from eggs via tadpoles, their ground actions are sliding—they are collecting materials to the aquarium and fiddling with the aquarium. Though the *play mood* is characterised by devotion, a shift in the *play mood* is occasionally happening when the children discover something new in the aquarium, for instance, when the aquarium suddenly is full of tadpoles.

Taken together, the *slow science* situation engages the children by mutually involving play and learning. The open-endedness of the playful situation invites all kinds of materials, questions, actions and people into the situation. Even though the *slow science* situation often is instructive and depends on ECEC staff's planning, overview and actions, the playfulness of the situation creates room for involving children's ideas.

2.4.3 Spontaneous Science

Spontaneous science is most often child-centred situations in the everyday life of ECEC centres when children use different *play media* (Skovbjerg, 2021), and it develops in situations where science and play are entangled (Ahrenkiel et al., 2023a).

From the data collected in *Science and Play*, this key episode was identified as a moment while playing where children were engaged in *spontaneous science* situations:

> *The children have made extractions from plants. There are many cups and bottles with liquid in different colours depending on which plants the children used. Now it is time to clean up. While Lene and some children are carrying cups and bottles to the sink, Emilia drags a pencil over the side of a bottle. A sound occurs. She tries again and again, and other children want to try. Lene asks: 'How come there are different sounds when the bottles are alike?' The children come up with different explanations: 'It is due to the different colours. No, no, it is because there is not the same amount of liquid in them', says another. 'It is because the liquid is higher up in some of them', tries a third child. 'Can we examine it?', asks Lene. They find more cups and bottles and compare cups and bottles with different colours, different liquid heights and so on. (Fieldnote)*

The situation is initiated by an action, Emilia, one of the children, performs. She drags a pencil over the side of a bottle and Lene, an ECEC staff member, reacts to the child's action, and the situation identifies as a *spontaneous science* situation, where Emilia and the other children are actively experimenting and undergo concrete experiences (Kolb, 2014) with the phenomenon of sound. The children's engagement with the sound of different bottles and cups are supported by the ECEC staff, who have an attention on questions that can lead to reflective observation and emerging hypothesis on the phenomenon of sound. By asking the question, '*How come there are different sounds when the bottles are alike*', the children's attention is directed towards an investigative play of 'what, if…, when…'. Reflective observations have the potential to identify differences, and the differences here on the quantity of liquid in the bottles are typically simple and clear in *spontaneous science.* An abstract conceptualisation (Kolb, 2014) of, for example, resonance, waves, tones etc. are not made explicit. However, the situation holds the potential to support such abstract conceptualisations (Kolb, 2014), especially when the ECEC staff are engaged in the situation, and omitting these concepts might be a result of the ECEC staff's professional judgement on what concepts the children are capable of understanding (given their age) and able to actively experiment with to have concrete experiences.

This *spontaneous science* situation emerges in the transition from a planned activity to a tidy-up situation. The shift is characterised by one child's play and investigation. The occurrence of new sounds created a sensory moment of surprise. The unpredicted action makes Lene react and ask an open-ended question that leads the children to playful experimentation characterised by the play mood devotion. The ground play action is sliding, when the children use pencils, bottles and cups with different quantities of liquid to make sounds. The children are focused and are using their hands and performing repetitive moves to create sounds and playful rhythms in their experimentation.

Broadly, play and investigation goes hand in hand and mutually support and potentially develop the situation.

2.5 Overview of the Key Episodes

In this section, we present an overview of the key episodes in the chapter. The overview relates the key episodes to the three science approaches, the four phases of Kolb's experiential learning theory, phenomena, central concepts and play moods (see Table 2.1 below).

Across the three key episodes, the most prominent play mood is devotion. The show science situation is planned for the children to have the possibility to play with flashlights and balls in a dark room and results in a play mood of intensity. It is worth noticing how materials in the learning environment support children's active

Table 2.1 Overview of the science approaches in an analytical perspective

Science approach	Show science	Slow science	Spontaneous science
Key episode	Solar eclipse	From egg to frog	Bottles with liquid
Phenomenon	Solar eclipse	Development from egg to frog	Sound
Concepts	Light, shadow, distance and orbit	Eggs, tadpoles, froglets, frogs, gills, lungs, metamorphosis and habitat	Resonance, frequency, sound wave and keynote
Active experimentation	Children and ECEC staff play with balls and flashlights	Children explore the pond with nets and buckets	Children use a pencil to explore sound from bottles and cups with liquid
Concrete experience	Children gain concrete experiences with light and shadow	By touching and investigating the egg, children gain concrete experiences of how it feels and what it resembles	Children gain concrete experiences with sound
Reflective observation	The ECEC staff asks open-ended questions about how the distance between the flashlight and the ball and are connected to the shadow	By taking some of the frog eggs back home, the ECEC staff creates an observational setting for seeing the development and for children to ask questions	ECEC staff asks questions on the relationship between the quantity of liquid and the sound
Abstract conceptualisation	The concepts of light and shadow are introduced as well as the concept of orbit	The concept of metamorphosis (Not introduced in the key episode)	The concepts of resonance and frequency (Not present in the key episode)
Play moods	Devotion and intensity with a moment of euphoric	Devotion with a moment of euphoric	Devotion

experimentation and concrete experiences and how materials support different play moods.

All three key episodes are based on a scientific phenomenon which appear to the children (solar eclipse, development from egg to adult and sound). However, the phenomena can be investigated in different ways. The solar eclipse is difficult to investigate, but the ECEC staff make it possible for the children to investigate with materials representing the Sun (flashlights) and the Moon (balls). This points to an awareness of how to involve children in show science, so they have possibilities for active experimentation and concrete experiences. In contrast, bottles with water are easy to investigate and offers many possibilities for active experimentation and concrete experiences for the children. It is characteristic of many spontaneous science that the situations are easier to investigate and play with.

In all three key episodes, children experience a phenomenon, and it requires a language to talk about this phenomenon. Here, the ECEC staff have the possibility of providing scientific concepts to the children as a natural part of the dialogue during their play and investigation of the phenomenon.

According to Kolb, there is a need both for conceptualising the active experimentation and concrete experiences and for concretising the reflective observation and abstract conceptualisation during playful science situations.

2.6 Discussion

In this chapter, we have unfolded science both as planned (show and slow) and as spontaneous situations. The three approaches cover a wide area of science situations, where each of them has special features.

Show science has the potential to introduce children to new phenomena and relate these with concepts of science. Learning science is a central part of the Danish ECEC curriculum, but there is no specific content within the ECEC curricula. Hence, we often see *show science* situations with a spectacular effect such as the creation of small volcanos with vinegar (acid) and soda (alkaline). To unfold the full potential in science, while it is necessary to relate the phenomena and concepts, it is also essential that children are given the opportunity to have concrete experiences with the phenomena through active experimentation (Kolb, 2014). *Show science* must therefore not only be 'to show' but also to sense the phenomena as we see in the key episode with solar eclipse. The phenomena must be concretised and not just explained in words and concepts (Vygotsky, 1986). Hence, *show science* needs to engage children with the phenomenon and ensure that children have concrete experiences. In the solar eclipse episode, we saw that the children were given the opportunity to experiment with light and shadow and have individual experiences instead of just being told about it. *Show science* without children's active experimentation gives experiences but not educative experiences (Dewey, 1913).

This is, of course, also true for *slow science.* However, *slow science* brings out different possibilities for science in that the changes happening with the phenomena often develop over a long period and are hard to grasp without repeated engagement. In the key episode with frog eggs, the metamorphosis from tadpoles to frogs happens within a relatively short period. In other *slow science* situations such as planting vegetables, the development is more continuous and therefore more difficult to see. A strength of *slow science* situations is the possibility for the children to make observations. Such observations may be modified by the ECEC staff to evolve from everyday observations towards more scientific observations (Eberbach & Crowley, 2009).

While active experimentation might be hard in *slow science* due to the slow development in the phenomenon, reflective observations can be underpinned and helped by simple documentation, for example, photos or drawings of the development. Different forms for documentation help children focus on the relevant features and ignore the irrelevant features and thereby improve the quality of the observations. Repeated observations of the slow development also help children establish expectations regarding what they observe (Howes, 2008).

The third category of science situation differs from the other two in that there is no preparation because it is spontaneous. *Spontaneous science* occurs when the ECEC staff recognise and capture science in the children's doings and play. While *slow science* can be good for developing children's observational skills, *spontaneous science* can develop the observational skills of the ECEC staff. The question is how to recognise what the children are doing that could relate to science.

Ahrenkiel et al. (2023b) have developed a set of science practices for children in ECEC and preschool settings. These science practices provide a picture of how to recognise science in children's play and doings. However, the hard part of capturing these moments is that often the ECEC staff are focused on other plans and need to put these plans on hold and go with the children taking the investigative approach. This is of course not always possible, but, when it is, there is a high motivation from the children, considering the idea of *spontaneous science* comes more often from them.

Until now, we have described *show, slow* and *spontaneous science* as separate ways of doing science; however, in practice, we find that the three approaches to science often intertwine, and that one type of science can lead to another.

In Fig. 2.2, we can distinguish between the ECEC staff and children in terms who initiates a science situation. We see that *spontaneous science* is more often initiated by children while *show science* is initiated by adults. However, a *spontaneous science* situation can quickly turn into a *show* or *slow science* situation and vice versa and thereby go along the other parameter, namely duration of the science situation. In the key episode with tadpoles, we do not know the intention of visiting the pond. If there was no intention other than to give the children a good experience with catching pond animals, the key episode can be seen as *spontaneous science* which then turns into *slow science.*

Across all three science approaches, play moods emerge. As we have seen, materials of various kinds frame the mind to different play moods and create room for active experimentation. Open-ended questions trigger investigative curiosity (Bonawitz et al., 2011), and the dialogic and open-ended-ness of play offer many

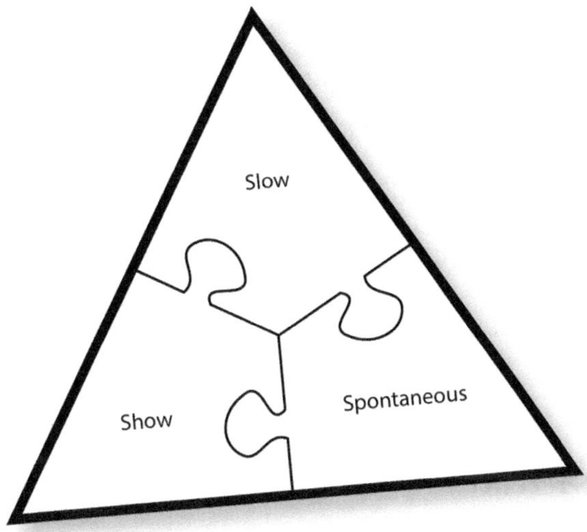

Fig. 2.2 A metaphorical model of the S'ence of early childhood science showing show, slow and spontaneous science as interlocking pieces in a puzzle

opportunities to make science become 'the material' in play in different ways. Following Bogost (2016), play is deliberately working with the materials you encounter, and it is therefore important to provide children with both concrete materials, language and actions to support the science approaches in playful ways as we have seen in the analysis.

2.7 Conclusion

In this chapter, we have shown that learning science appears in different forms and with different qualities. Together, *show, slow* and *spontaneous science* cover a large field of what early years science is. Being aware of the different approaches to science and their individual qualities, the ECEC staff have the advantage of seeing beforehand if a science situation is predominantly inclined towards one approach and can accordingly expand the area of science activities to cover all three approaches more evenly.

Drawing on Kolb (2014), we argue that the ECEC-staff initiated science situations (mostly *show* and *slow science*) entail more than introducing a phenomenon or concept to children. The science activities need to open up for children's play and active experimentation so that they have concrete experiences. This means, for example, that *show science* must show the children how they can experiment with the shown phenomenon or concept. Hence, it is crucial to pick carefully both content and materials with which children can experiment. This also calls for an awareness from the ECEC staff on the relatedness between science and children's play. In this chapter we have outlined that science activities also holds the potential for addressing play moods and that play often contains a science content.

Another conclusion of this chapter is that science is present in many situations. It is not always necessary to have extensive preparation and special materials for doing science. Often, it is to grasp the science in children's play. To do so requires the ability of the ECEC staff to recognise science in play and to participate in the play, thereby introducing science concepts as a natural part if the dialogue with children. This might not work every time, but, with practice, it will work more and more, connecting science and play in a meaningful way. Such an approach would also be fruitful in a primary school setting for maintaining or attaching positive moods towards school science.

References

Ahrenkiel, L., Jørgensen, H. H., & Petersen, M. R. (2023a). *Sciencepraksisser og legekvaliteter – redskaber til legende og undersøgende tilgange til science i dagtilbud* [Science practices and play qualities – tools for playful and inquiring approaches to science in daycare and preschool]. UCL, Denmark Online: https://www.ucl.dk/science-og-leg

Ahrenkiel, L., Petersen, M. R., & Jørgensen, H. H. (2023b). Science practices as a tool for spotting and supporting children's investigative actions in Early Childhood Education and Care (ECEC). *Journal of Emergent Science, 25*(2), 5–11.

Akker, J. V. D., Gravemeijer, K., McKenney, S., & Nieveen, N. (Eds.). (2006). *Educational design research.* Routledge.

Bogost, I. (2016). *Play anything: The pleasure of limits, the uses of boredom, and the secret of games.* Basic Books.

Bonawitz, E., Shafto, P., Gweon, H., Goodman, N. D., Spelke, E., & Schulz, L. (2011). The double-edged sword of pedagogy: Instruction limits spontaneous exploration and discovery. *Cognition, 120*(3), 322–330. https://doi.org/10.1016/j.cognition.2010.10.001

Braun, V., & Clarke, V. (2006). Using thematic analysis in psychology. *Qualitative Research in Psychology, 3*(2), 77–101. https://doi.org/10.1191/1478088706qp063oa

Dewey, J. (1913). *Interest and effort in education.* Franklin Classics Trade Press.

Eberbach, C., & Crowley, K. (2009). From everyday to scientific observation: How children learn to observe the biologist's world. *Review of Educational Research, 79*(1), 39–68.

Emerson, R. M., Fretz, R. I., & Shaw, L. L. (2011). *Writing ethnographic fieldnotes* (2nd ed.). The University of Chicago Press.

Eshach, H., & Fried, M. N. (2005). Should science be taught in early childhood? *Journal of Science Education and Technology, 14*, 315–336.

Fleer, M. (2009). Understanding the dialectical relations between everyday concepts and scientific concepts within play-based programs. *Research in Science Education, 39*, 281–306.

Fleer, M. (2022). Balancing play and science learning: Developing children's scientific learning in the classroom through imaginary play. *Journal of Emergent Science, 22*, 13–20.

Gulløv, E., & Skreland, L. L. (2016). Ethnographic studies of young children. In J. Prior & J. Van Herwegen (Eds.), *Practical research with children* (pp. 127−144). Routledge.

Hansson, L., Leden, L., & Thulin, S. (2021). Nature of science in early years science teaching. *European Early Childhood Education Research Journal, 29*(5), 795–807.

Howes, E. V. (2008). Educative experiences and early childhood science education: A Deweyan perspective on learning to observe. *Teaching and Teacher Education, 24*(3), 536–549. https://doi.org/10.1016/j.tate.2007.03.006

Jørgensen, H. H., & Skovbjerg, H. M. (2024). Play qualities: Playful actions in learning processes in teacher education. *Social Sciences & Humanities Open, 9*.

Kolb, D. A. (2014). *Experiential learning: Experience as the source of learning and development*. FT Press.

Larimore, R. A. (2020). Preschool science education: A vision for the future. *Early Childhood Education Journal, 48*(6), 703–714.

Ministry of Children and Education. (2020). The strengthened pedagogical curriculum, Framework and content. ISBN. 978-87-7182-366-0. Available from https://emu.dk/sites/default/files/2021-03/8077%20SPL%20Hovedpublikation_UK_WEB%20FINAL-a.pdf

Skovbjerg, H. M. (2021) *On play*. Samfundslitteratur.

Sonneborn, R. (1982). *Hvem spiser solen?* [Who eats the Sun?]. Carlsen, København.

Vygotsky, L. S. (1986). *Thought and language* (Alex Kozulin, Trans. & Ed). MIT Press.

Zahavi, D. (2018). *Phenomenology: The basics*. Routledge.

Chapter 3
A Formative Assessment Sequence to Support Children's Learning of Scientific Concepts with Images

Jennifer Yeo and Kim Chwee Daniel Tan

Abstract This chapter introduces a formative assessment (FA) sequence aimed at helping teachers support primary school students in constructing representations to learn scientific concepts. Recognising the challenges teachers face in working with students' conceptual learning, the FA sequence provides a structured method for teachers to work with student-generated representations. Drawing on the theory of multimodality, the sequence encourages teachers to engage with students' visual representations, helping them connect these ideas to formal scientific concepts. The key stages of the FA sequence—explore, work-on, and review—demonstrate how they assist teachers in facilitating students' understanding of complex ideas. By focusing on interpreting and extending students' visual representations, this approach bridges the gap between students' initial ideas and formal scientific knowledge. A case example is used to highlight the practical application of the FA sequence, underscoring its value in fostering conceptual, representational, and epistemic growth in science concept learning. This chapter demonstrates the importance of guided support in representation construction for effective science education.

3.1 Introduction

The challenges that primary school students face in learning scientific concepts have been well-documented since the 1980s (e.g., Driver et al., 1985; Posner et al., 1982). For example, it has been reported that children found it difficult to differentiate between the concepts of "temperature" and "heat" (e.g., Paik et al., 2007; Thomaz et al., 1995). While traditional conceptual learning theories have attributed these challenges to the alternative ideas conceived from one's everyday experiences with

J. Yeo (✉)
Singapore University of Social Sciences, Singapore, Singapore
e-mail: jenniferyeoac@suss.edu.sg

K. C. D. Tan
National Institute of Education, Nanyang Technological University, Singapore, Singapore
e-mail: daniel.tan@nie.edu.sg

S. Earle et al. (eds.), *Primary Science Learning for Children, Teachers, and Communities*, https://doi.org/10.1007/978-981-95-0561-6_3

the world (Duit & Treagust, 1995), contemporary science educators point to the **technicality, abstractness** and multimodal nature of the scientific language as the reason (Hamnell-Pamment, 2024). It was found that students have problems comprehending the concept of "temperature" as an ordinal measure of the degree of hotness; instead they thought of it as a material property or an instrument for measuring heat (Park et al., 2007). To address the difficulty of learning scientific concepts, images are commonly used to help students **visualise and concretise** abstract ideas.

The utilisation of images as a tool for teaching and learning is not a novel strategy as it is common for science teachers to include pictures and videos in their instructional resources to better communicate scientific ideas (e.g., Lee & Jones, 2017). In more recent years, there has been a shift from presenting images to aid interpretation of scientific ideas to getting children to create representations as means to construct meanings for these scientific ideas. Examples of image-constructing methods include Representation Construction Approach (RCA, Tytler et al., 2013), Ainsworth et al.'s (2011) Drawing to Learn Science and Image-to-writing (I2W) approach (Yeo et al., 2020). While representations have significant potential support concept learning (Pappas & Varelas, 2009), studies (e.g., Hubber et al., 2010) suggest that students need to be supported in the process to build representation and meta-representational competence (Britsch, 2013). These include discussing conventions involved in the generation of representations and using representations as tools for learning.

While science educators are generally adept at instructing children with images, challenges arise when it comes to engaging with the images produced by the students themselves, in terms of grasping their intended meanings and facilitating the transition to canonical ideas and representations (Prain & Tytler, 2013). Teachers mistakenly assume that students can naturally connect meanings and different forms of representation on their own, leading them to overlook the necessary support needed to help students comprehend these visual aids (Eilam, 2012). Similar findings were also reported in the first cycle of the present study which designed intervention to support students' conceptual learning in an I2W approach. In addition, our teachers struggled to work with students' representations (images or text) even when recognising potential misconceptions. They were unsure how to identify or address these representations, resulting in missed opportunities to correct misconceptions (Yeo et al., 2020). Prain and Tytler (2013) also noted that teachers faced new demands on their teaching skills when using RCA in science. Similarly, Gray and Rogan-Klyve (2018) found that although teachers understood metamodeling ideas, they failed to explicitly discuss them with students, despite being presented with clear opportunities. It is in the context of these challenges that a formative assessment (FA) sequence, explore—work-on—review, was introduced into the image-to-writing (I2W) approach to guide teachers in representation constructing activities.

This chapter will be based on the I2W approach introduced to two local primary schools to support science concept learning. The approach is a representation construction learning method that engages primary school students in constructing and working with visual representations to think about concepts before the formal language of science is introduced. The FA sequence was introduced to guide teachers in facilitating the interpretation and extension of students' ideas, as represented in their images, towards the scientific concepts outlined in the syllabus' learning outcomes. In this chapter, our aim is to introduce the three-stage FA sequence and illustrate it with one case example to allow readers to have a deeper understanding of the intricacies of supporting students in constructing and refining their representations to develop the conceptual, representational and epistemic aspects of science concept learning. The analysis of the enactment of I2W was focused on the following research questions:

1. When was the FA sequence used during I2W activity?
2. What was the purpose of the FA sequence during I2W activity?
3. What were the pedagogical strategies used during the FA sequence?

3.2 Conceptual Framework

The theory of multimodality (Kress & van Leeuwen, 2006) informs this study. The theory posits that different types of representations have different affordances, and the choice of representations selected for meaning-making is motivated (intentional) and culturally and socially shaped. We further employ multimodal analysis to identify the pedagogical strategies and actions used in the FA sequence.

3.2.1 Defining Scientific Concepts and Science Concept Learning

Taking a multimodal perspective, a scientific concept can be described as "a network of semantic meanings, assembled across multiple modes of representations" (Tang & Tan, 2017, p. 22). The concept of "temperature" (i.e., temperature is a measure of the degree of hotness) for example, is defined in terms of technical words (i.e., measure, degree, hotness) through grammatical resources, conceptualised as an ordinal quantity (in terms of relative hotness), and represented by numbers and symbols (e.g., 50 °C) with each mode of representation (e.g., hotness vs 50 °C) serving a particular epistemic purpose. As the example shows, concepts are often defined generally and without context, making their representations unfamiliar to everyday situations. This abstraction and technicality are frequently blamed for students' difficulties in learning scientific concepts.

The semantic and semiotic characteristics of a scientific concept mean that concept learning must necessarily involve developing the conceptual and representational aspects of the idea. Furthermore, the different semiotic/representational types for the same concept but serving different purposes signal the epistemic aspect of concept learning. Learning science concepts should thus involve **making sense of the ideational meaning (conceptual) inscribed in the institutionalised representations, the affordance of the chosen representations for the motivated purpose**

(representational), and the motivation for the choice of the representations (epistemic). These three aspects (conceptual, representational and epistemic) form the fundamental foci of the I2W approach as well as the FA sequence.

3.2.2 *Science Concept Learning: What It Entails*

We adopt the view that learning is a transformation and transduction of semiotic forms (Kress, 2010). Kress (2010) defines "transformation" as the process of re-ordering the elements or entities in a text or artefact in the same mode, whether from one genre to another, or from one language to another. Applied to science learning, it involves selecting, ordering and re-ordering representations within the same mode (e.g., re-writing or recapping a textual definition) that learners consider apt and aligned with their interest and motive. Transduction (Kress, 2010) refers to remaking meaning across modes (e.g., from a table representation of data to a graphical representation). The concept acknowledges that meaning expressed in one mode is never identical to that in another, even for the same idea. Thus, transduction involves a meaning-making process where learners must interpret the meaning as presented in one mode and reconstruct it using the affordances of another.

Applying the concepts of **transformation and transduction** to conceptual learning, we note that while scientific concepts are generally abstract, their origins can be traced to the physical world. The concept of heat, for example, is related to changes in temperature. The concept of temperature, on the other hand, is related to the sensory touch of hot and cold. Yet, these ideas are represented in symbolic forms like words and numbers that bear little or no resemblance to their physical manifestation.

Philosophers of science, such as Nersessian (1992) and Gooding (2004), have illustrated how scientists navigate a series of transformations and transductions of semiotic modes. Latour (1999) characterises these processes as "representational passes," which are essential for developing institutionalized and abstract scientific concepts, laws and theories, like electromagnetism. Similarly, research by Yeo and Gilbert (2014, 2022) indicates that students also undergo representational passes across various modalities when constructing scientific explanations for real-world phenomena. These processes of transformation and transduction suggest an abstraction whereby conceptual, representational, and epistemological meanings are embedded within representations. Consequently, effective science learning should involve unpacking these meanings.

The I2W approach (Yeo et al., 2020), which forms the context of this chapter, is modelled after this transduction and transformation process. It comprises three main stages (refer to Fig. 3.1): (1) exploring a phenomenon, (2) creating and transforming images, and (3) transducing images to writing. An example of the I2W approach for the concept of temperature involves primary students exploring different methods of measuring temperature of cups of water such as using the sense of touch, thermal imagery and thermometer (Stage 1). Each method realises an aspect of the concept,

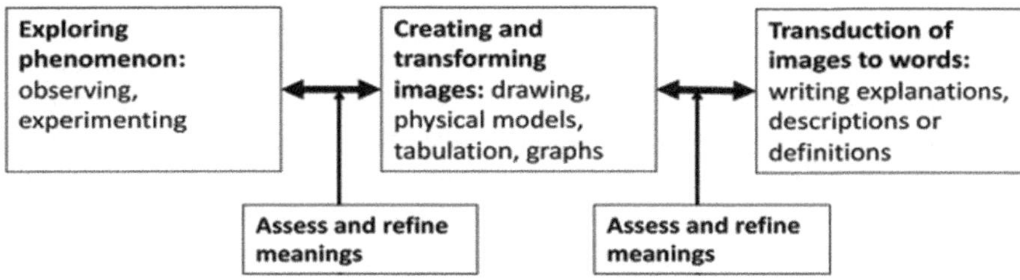

Fig. 3.1 The process of image-to-writing approach (Yeo et al., 2020)

from the meanings of "degree" and "hotness" from the ordinal arrangement of the cups of water based on a sense of touch to the purpose of a scale for consistency in describing and using numbers as "measure" of hotness due to its affordance for further mathematical manipulation. Students generate temperature scales from these various methods and work on them to interpret their meanings along the three dimensions (Stages 2 and 3). As mentioned earlier, while the generation of images opens up opportunities for discussion of concepts, teachers and students often find it difficult to engage in such discourse. A formative assessment sequence introduced to the "assess and refine meanings" stage is thus designed to address this issue.

3.3 A Formative Assessment (FA) Sequence for Working with Students' Images

The fundamental principle of FA is that student learning can be better facilitated if the teacher knows where students are along the paths toward specific goals (Black & Wiliam, 2009). This ensures that learning is facilitated in the students' zone of proximal development (ZPD, Vygotsky, 1978). The essential processes of FA are (1) establishing where the students are in their learning, (2) establishing where they are going, and (3) establishing what needs to be done to get them there (Black & Wiliam, 2009). A formative assessment structure "Explore—Work-on—Review", adapted from Mortimer and Scott's (2003) teaching sequence, guides teachers in performing FA in the I2W approach. Mortimer and Scott (2003) demonstrated how this pedagogical pattern worked as teachers orchestrated talk in supporting students to shift their conception between everyday-empirical experiences and theoretical-scientific ideas.

(1) Explore

The "Explore" phase seeks to establish where the students are in their learning. The theory of multimodality informs us that the images, be it a drawing, chart, graph or physical model, that students create are evidence of their cognitive processes. In the context of I2W approach, students are engaged in creating and interpreting images (e.g., a touch scale) and other signs from concrete/ everyday experiences (e.g., using fingers to sense hotness of water). At this stage, these images should not be evaluated based on the language system of science or the teacher's. Through an interactive discourse, the child should be allowed to explain the meaning he/she is making with the representation (conceptual), the choice of representations for the purpose (epistemic) and the affordance the representations have (representational), which will enable the teacher to better assess the three aspects of meanings the child is making with the representation(s).

(2) Work-on

Having grasped an understanding of students' meaning-making, the work-on phase entails working on students' meanings to bring them to more acceptable forms of scientific representations, according to the national science syllabus. The teacher needs to analyse the students' learning along the cognitive, representational, and epistemic aspects of concept learning (e.g., the aspects of temperature as described in previous sections) from what was gathered from the "explore" phase.

He/she then needs to establish how to achieve the learning objectives by working with the students' ideas. This is probably the most important but least understood process. Khan (2011) found that teachers rarely refined students' constructed models during model-based teaching. Similarly, Waldrip and Prain (2013) reported that teachers faced difficulties in guiding students from self-generated representations to scientific ones, highlighting the challenges in facilitating deeper understanding of the nature of representations. Yet, a meta-synthesis of research on learning with diagrams (Tippett, 2016) highlights importance of a teacher's role in mediating discussions on the cognitive strategies (conceptual), use of representations (representational), and conventions and traditions of scientific representations (epistemic) to build their representations and meta-representational competences for concept learning.

(3) Review

Mortimer and Scott (2003) observed that the "review" phase plays an important role in the iterative rhythm of classroom talk. It helps to "draw the line" (p. 71) under all interactions and makes a statement of where students ought to go in terms of the learning objective. In this phase, teachers review and summarise progress and make a statement of what students should know at a particular juncture of learning.

As novices in science, elementary students may not have the representational resources that scientists or science teachers possess. The drawings they produce, the physical model they build, or the words they use reflect their system of language. If learning science is appropriating the language of science (Lemke, 1990), students need to be engaged in thinking and reasoning about the aptness of the signs they produce to represent their meanings and compare them with those used in school science. Teachers play a significant role here in introducing new requisite signs and representations (e.g., scientific vocabulary, scientific drawing) and guide them in coordinating the different representations in the transduction of one form of representation to another (Tytler et al., 2013).

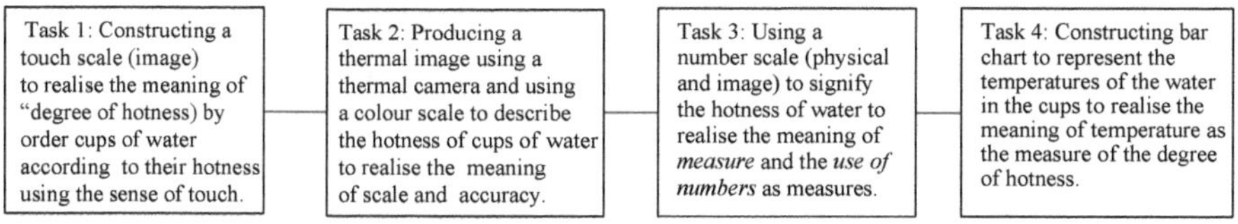

Fig. 3.2 Sequence of image-to-writing activities for the concept of temperature

3.4 Research Context

Design research was employed as the methodology for this study. Three research cycles were carried out, with the results of each informing the design and implementation of the next cycle. In this chapter, we focused on our findings from Research Cycle 2. The FA sequence is taken from the activity to develop an understanding of the concept of "temperature". This activity consisted of four image-constructing tasks that involved the exploration of different representational modes (i.e., touch, colour, number) for describing the degree of hotness of five cups of water (see Fig. 3.2.).

The illustrated case example is based on Task 2, conducted in one 60-min lesson. From Task 1, students had learnt four key ideas:

(1) touch is one modality for telling how hot something is,
(2) the touch scale (refer to Fig. 3.2) provides a means to "rank" the varying degree of hotness,
(3) temperature is the terminology to refer to the "degree of hotness",
(4) words like hot/cold are not useful for describing the degree of hotness.

In Task 2, students were to learn that a colour scale can be used to describe the degree of hotness of an object and that thermal imaging is more reliable compared to the use of touch.

3.5 Enactment of FA Sequence in I2W Approach

We found three key instances during the stages of the I2W approach where FA was used for the development of the concept of temperature. These three instances took place at the following junctures: (1) "pre-image-constructing task" to ensure that the students had the prior knowledge required before proceeding with the image-constructing activity; (2) "post-image-constructing task" to find out the meanings students had made during the activity, and (3) "meaning extension" to expand on the ideas students had made during the activity. We illustrate the three instances with Task 2, with excerpts taken to illustrate them. It is to note that various students' talk is indicated as "Student" under the column "Speaker" unless there is a specific reference to the student's name by the teacher. The specific name of the student is replaced with a code (e.g., Student 1, Student A).

3.5.1 Pre-task 2 FA: Before We Start Today's Experiment, There Are a Few Words You Must be Very Familiar

In this excerpt, we see the teacher using an FA sequence to review the concept of "temperature" in which he includes the three phases important to formative assessment: explore, work-on and review, to highlight two key concepts "scale" and "temperature" needed for Task 2 (Table 3.1).

(1) Explore

In this first stage of the FA sequence, the teacher is trying to ascertain the prior knowledge that the students have. With the context-general question "What is temperature?" in Turn 1, the teacher identifies two related aspects of the concept that students are struggling with—the epistemic use of "hot" and "cold" (Turns 2 and 8) in the definition of temperature and the conceptual meaning of "scale" in ordering levels of hotness (Turn 10).

(2) Work-On

Having identified the epistemic and conceptual challenges of his students, the teacher embarks on "working-on" them. In both cases, the touch scale (refer to image in Turn 11) previously constructed by the students to describe the hotness of the cups of water plays an important role.

Making specific reference to the technical terms ("least hot" to "hottest") on the touch scale the teacher reminds the students of the agreed use of "hotness" and not "coldness" to describe the temperature of water in the previous lesson (Turn 11). He further relates how such "institutionalised" technical terminology can allow weather to be described in their everyday life (Turn 19). In so doing, the epistemic function of "hotness" that is inscribed in the touch scale is realised and why "hotness" is used to describe temperature rather than "coldness" is explained.

In the same excerpt, we also see the teacher using the touch scale to realise the conceptual meaning of "scale", another technical term associated with the concept of "temperature". In Turn 21, the teacher first draws an abstraction of the "hotness" scale and then refers to the physical touch scale that the student has constructed. In so doing, he concretises the meaning of the word "scale" with an abstracted and general image of a scale representing the comparative degree of "hotness" along the horizontal arrow and a more contextualised image of the touch scale showing images of the cups of water ranked.

(3) Review

At the end of this "working-on" stage, the teacher "reviews" by way of checking if students recognise the two key concepts that are discussed (Turns 23 to 28). This FA sequence thus ensured that the representational and epistemic aspects of the concept of "temperature" are revisited for use in Task 2.

In summary, Excerpt 1 shows how the FA sequence guides the teacher to find out, work-on and review the issues students have with the concept of temperature in

Table 3.1 Excerpt 1

Turn	Speaker	Utterance	FA sequence
1	Teacher	… So in the previous lesson, we spoke about temperature, okay. And can we say together what is temperature? What is temperature?	Exploring students' conceptual understanding of "temperature"
2	Student	To measure heat!	
3	Teacher	Uh huh. Student 1	
4	Student 1	Feeling, feeling	
5	Teacher	Feeling, okay. Erm, Student 2?	
6	Student 2	Measure the degree of hotness	
7	Teacher	Measure the degree of hotness	
8	Student	Coldness	
9	Teacher	No, no, no. That time like I mentioned to you. Remember the scale that we did?	Working on the epistemic use of "hot" and "cold"
10	Student	What scale?	
11	Teacher	Ah, what scale? Okay good. Remember this one? Remember this? What did we use?  *(Teacher takes out the touch scale and gestures to the words "least hot" and "hottest")* Least hot to hottest. Why am I using such words? Least hot to hottest? What am I doing? I am actually measuring the degree of hotness. If I want to measure the degree of coldness, I wouldn't use ...	
12	Student	Just say least coldest	
13	Teacher	Ah, I wouldn't right? Yes, Student 3?	
14	Student 3	How come you cannot say the hot and cold?	
15	Teacher	You see. Previously, before we come to this lesson, we can only tell hot or cold. But how hot, how cold?	
16	Student	Then how hot or how cold?	
17	Teacher	How hot or how cold? I wouldn't know right? … That time I remembered you mentioned the weather station, so people want to tell today's weather is hot but how hot?	
18	Student	Very hot!	

(continued)

Table 3.1 (continued)

Turn	Speaker	Utterance	FA sequence
19	Teacher	Ah, so before you come to our class, you will probably say it is very hot. But now, with our lesson, you'll probably have learnt this (teachers takes out the touch scale again). If let's say you can show people, then you'll probably have a scale from least hot to hottest (pointing to the words, "least hot" and "hottest") Just now, one of you asked what is a scale right? …	
		… [classroom management matters]	
21	Teacher	So now from least hot to the hottest 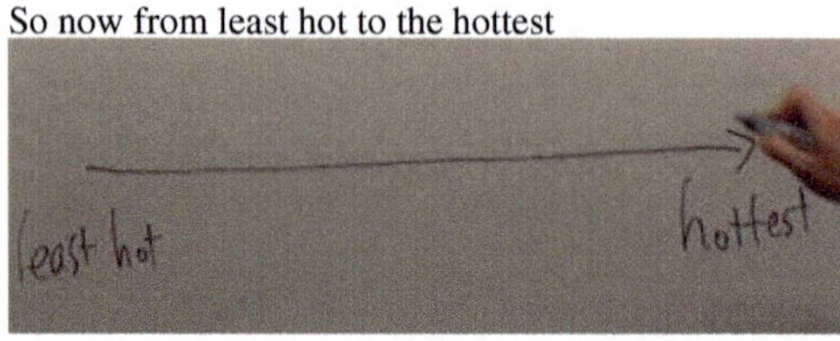 *(Teacher draws on the board.)* This is what we call a scale. A scale can tell you more details, correct or not? … Now if I can tell, I can show people this (refering to the touch scale teacher is holding), so I now tell you, today's weather is like this, you probably would know it's closer to the [hottest]. Okay, or closer to least hot, you all decide. But is this good enough?	Working on the conceptual and epistemic meaning of "scale"
22	Students	No!	
23	Teacher	Ah no, right? Why? That time we use what? The touch. Remember we used the touch scale. That's why I add the word "touch scale" on that piece of template right? Meaning you are using your finger, you sense of touch as an apparatus to feel it and then you group accordingly the six cups. …	
		Today, then I am going to get you to do the thermal imager. So today, okay, before we start today's experiment, there are a few words you must be vary familiar. So, I'm going to write down on the board, what are those words. …	Reviewing the two key terms discussed and needed for the next task
24	Student	Scale!	
25	Teacher	Scale. Okay scale. One more?	
26	Student	The touch	
27	Teacher	The touch scale. Just now, I mentioned the word with the letter T	
28	Student	Touch	
29	Student	Temperature	
30	Teacher	Temperature okay	

a systematic manner. In addition, we also observe the key role played by the touch scale in realising the epistemic function of key technical terminologies. The touch scale is an abstraction of the students' concrete experiences with temperature, rich in the conceptual, epistemic and representational meanings that are inscribed in the previous activity. Therefore, by referring to the specific experiences and modalities of the touch scale, the conceptual and epistemic meanings inscribed in them are brought to the fore.

3.5.2 Post-task 2 FA: Can I Have Everyone to Look at Your Colour Scale?

After reviewing the key concepts needed for Task 2, the students are instructed to capture a thermal image of the six cups of water with a thermal camera. They then match the thermal image of each cup to the colour scale and indicated their positions on the scale. The purpose of the activity is to introduce the idea of a more precise scale. Excerpt 2 shows the discussion that follows immediately after students have completed Task 2 (Table 3.2).

(1) Explore

With the teacher's initiating question (Turn 65), students' understanding of the epistemic purpose of the colour scale is made visible. While none of the responses (Turns 66 – 72) gives the expected generalised answer that the colour scale tells "the degree of hotness", one response, that it measures "heat", stood out as a common misconception, which the teacher subsequently addressed.

(2) Work-On

The teacher reinforces prior learning on temperature ("degree of hotness" from Turn 75) by drawing an abstracted touch scale and referring back to the tactile experience of feeling water temperature with a finger. Through this transduction, the teacher links the general definition of temperature to the abstract image and tactile experiences, helping students construct the concept of "degree of hotness" from these representational stages (Turns 75–78). By drawing parallel between the touch scale and Task 2 ("today, instead of using touch scale, we ... use the colour scale" in Turn 79), the students were thus able to identify the use of colour scale as similarly the means to measure the "degree of hotness" (Turn 80).

(3) Review

Linking the colour scale with temperature, the teacher reiterated that temperature is "the degree of hotness" (Turn 75) and the colour scale "tells us the degree of hotness" (Turn 81).

In summary, the FA sequence in Excerpt 2 identifies and addresses a common misconception about temperature. Working on the misconception, we see a series

Table 3.2 Excerpt 2

Turn	Speaker	Utterance	FA sequence
65	Teacher	Okay, we are using the colour scale to do what?	Explore the epistemic meaning of colour scale
66	Student	Measure the amount of heat that we saw on the thermal imager	
67	Teacher	Measure the amount? What else? What else is it doing? What are we actually doing with the colour scale? Is it to measure the amount of heat?	
68	Student	No	
69	Teacher	No, what do you think it is? What do you think it is?	
70	Student	The lightest colour is cold and … the brighter colour is hot	
71	Teacher	Okay, so the colour scale is actually telling us what?	
72	Student	That it's hot	
73	Teacher	… What have we just learnt just now? The degree of?	
74	Student	Hotness!	
75	Teacher	Hotness right? Temperature. Just now, remember, boys, we talked about temperature as the degree of hotness. (teacher writes on the board "Temperature—degree of hotness") And then, we use the **touch scale**. Okay, I'm going to write down what we did earlier in the previous lesson, so that you can link to what we did. (teacher draws an abstracted touch scale on the board). 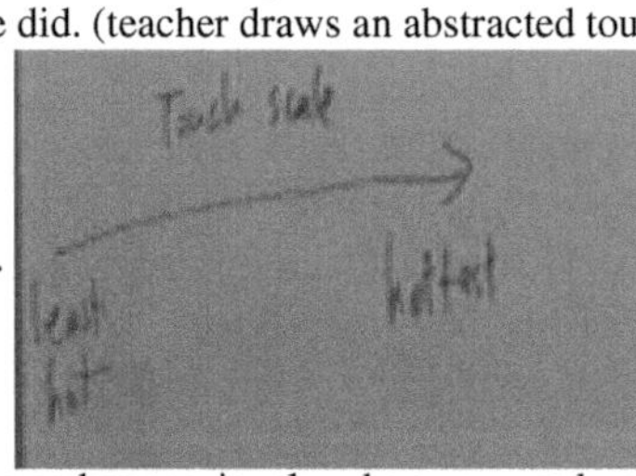 So, you remember previously when you used your finger, does it tell us the amount of heat?	Working on epistemic meaning of colour scale for measuring degree of hotness, and not amount of heat
76	Student	No!	
77	Teacher	No right? It tells us what?	
78	Student	Temperature	
79	Teacher	Temperature, okay. So today, instead of using the touch scale, we are going to use the colour scale to what? Tell us the?	Reviewing the epistemic purpose of colour scale
80	Student	Degree!	
81	Teacher	Degree of [hotness]. (teacher writes on the board) Okay, to tell us **the degree of hotness**. …	

of representational passes used to first arrive at how the touch scale and then the colour scale tells "degree of hotness" and not "degree of heat". Following this, the teacher continues unpacking the meaning of the various colours in the colour scale and relates them to the degree of hotness to ensure that the students can relate the arrangement of cups of water to the colour scale and the concept of temperature.

After confirming that students understand the colour scale, the teacher introduces the concept of "accuracy" by comparing the touch and colour scales in the next excerpt.

3.5.3 *Meaning Extension: Why Do You Think You Actually Prefer Using the Colour Scale?*

Excerpt 3 illustrates an interaction between the teacher and students as they discuss their preferences for different temperature scales and the reasons behind their choices. The purpose of these questions is to facilitate a comparison between the two scales, ultimately leading to the conclusion that the colour scale is more accurate than the touch scale (Table 3.3).

(1) Explore

In Excerpt 3, we see an exploration of students' preference between using the touch and colour scales for measuring the degree of hotness and why. Ideas from different students are sought, and there is no evaluation from the teacher. Mortimer and Scott (2003) refer to such interaction as being dialogic. At the end of the "explore" phase, it is clear that most students mention that they prefer the colour scale because it is "easier" to use, while a couple of students (e.g., Student 10) are not able to express themselves adequately.

(2) Work-On

Recognising the representational challenge that students are having in describing the comparison between the two scales. the teacher introduces the term "accurate" directly to them (Turn 174). While the strategy used by the teacher seems rather contrived, it might be difficult for students to arrive at the technical term "accuracy" on their own.

(2) Review

Finally, the teacher summed up Task 2 by authoritatively stating the two ideas that students need to know: (1) "a scale to tell us the degree of hotness of the cup(s of water), and "the colour scale is actually more accurate than using the touch (scale)".

In summary, Excerpt 3 highlights a distinct aspect of the FA sequence, featuring an extended interaction where the teacher invites various students to share their preferences without evaluating their responses in the explore phase. This dialogic approach (Mortimer & Scott, 2003) contrasts with other excerpts. It provides students a safe space to offer their ideas, and at the same time, allows the teacher to find out the meanings that students make. Additionally, the excerpt demonstrates how learning issues in a student-centered representation activity can be complemented by direct instruction. When students struggle to use correct scientific terminology to express their preference for the colour scale, the teacher introduces the term "accurate." As

Table 3.3 Excerpt 3

Turn	Speaker	Utterance	FA sequence
154	Teacher	Colour scale. Okay, now, Student 7 just now you would like to explain to us, why do you think you actually prefer using the colour scale?	Exploring students' comparison of the representational affordance of the colour scale vs the touch scale
159	Student 7	Because when you take the picture of the cups, there's the colour inside the water, so you can like, exactly, put the colours in the colour scale, 'cause it's the colours of the colours scale	
160	Teacher	Okay. So you can, it's easier for you to classify the cups of water according to the colour scale, is it?	
161	Student 7	Yes	
162	Teacher	Okay, Student 8, just now you also say you would like to use the colour scale. Okay, can you explain or tell us more. Why is it that your group prefers to use the colour scale than the touch scale?	
163	Student 8	()	
164	Teacher	Not sure. But you find that it's easier for you? Yes, Student 9?	
165	Student 9	Because … because with the thermal scale	
166	Teacher	Thermal scale?	
167	Student 9	Thermal imager. Then you can just see the colour of the cup and then you can just measure it	
168	Teacher	And you can just measure it? So between the colour scale and the touch scale. The colour scale and the touch scale. Yes, Student 10 you have something to say first?	
169	Student 10	Erm, the colour scale is easier because. Erm, if you measure with colour, it is even easier	
170	Teacher	It's even easier? So okay, you're trying to say something is it? That's it? Anything else? Okay, now so. Yes, Student 11?	
171	Student 11	()	
172	Teacher	Shhh	
173	Student 11	The colour scale is easier because, err, ()	

(continued)

Table 3.3 (continued)

Turn	Speaker	Utterance	FA sequence
174	Teacher	Erm, okay, now, ... Am I right to say that Student 10, you find that using the colour scale is actually more accurate instead of. Easier definitely, because you are using the iPad. ... No, because Student 10 started wanting to say one word that seems like, but it just couldn't come out from his mouth. So just guess and see perhaps it is what. Okay, so, you will see that using the colour scale. Will you agree with me that using the colour scale and also the thermal imager, it is actually more accurate?	Working on students' use of words (representational)
175	Students	Yes	
176	Teacher	Okay, for you to arrange the cups according to their degree of hotness. Rather than the touch right?	
	Students	(various students mumbling)	
177	Teacher	Touch is not accurate because?	
178	Student	Because sometimes your finger, may have only ()	
179	Teacher	Yah	
180	Student	Because your finger is not a good conductor	
181	Teacher	Oh finger is not a good conductor of heat. Okay. Alright, so in today's lesson. Shhh. I just want everybody to bring home one important idea first, okay. And this is what I want you to take note of. (teacher prepares the screen) This is the one I want everybody to take note of. Okay, so colour, in today's experiment, is actually used as a scale, okay, as a scale to tell us the degree of hotness of the cup. Okay, you are using the colour scale to actually tell us the degree of hotness. And I think we come across as a class and we agree that the colour scale is actually more accurate than using the touch	Reviewing the representational affordance of colour scale compared to touch scale

the word "accurate" is technical, it is unlikely students will have encountered it in their everyday life. As such, direct instruction becomes essential for understanding the conceptual meanings of scientific conventions and terminology.

3.6 Discussion

This chapter asserts that scientific concepts consist of networks of meanings across various modes of representation, encompassing conceptual, representational, and epistemic dimensions. Our multimodal perspective highlights the transformative processes in students' conceptual development, which can be challenging for both students and teachers. The FA sequence—explore, work-on, and review—supports teachers in assessing and addressing students' ideas within their Zone of Proximal Development (ZPD). The findings from the excerpts illustrate the FA sequence's utility and provide strategies to tackle challenges in learning scientific concepts.

3.6.1 Utility of FA Sequence

Our findings indicate that formative assessment is valuable both before and after a task. Scientific concepts are interconnected (e.g., "temperature" and "scale"), with each new concept building on previously learned ones (Bransford et al., 2000; Vosniadou, 2007). Mastery of earlier concepts is essential, as gaps can impede understanding of later ones. This is especially pertinent in representation-constructing activities like the I2W approach, where learning involves dialectical constructing and interpreting representations (Tippett, 2016). Students must understand the relationships between concepts and their dimensions to build new connections in different contexts.

The FA sequence proves useful in analysing students' activities by interpreting and extending their understanding. In representation-oriented tasks, students must make sense of the various representations they work with and connect them to previously learned concepts. However, understanding a concept in one context (e.g., touch scale) does not guarantee that students can transfer that understanding to a new context (e.g., colour scale), as illustrated in Excerpt 2 and reported in other studies (e.g., Eilam, 2012). Thus, the FA sequence supports students in interpreting the tasks they participate in and allows teachers to work within their ZPD. In the case study, the teacher was able to identify a common misconception and address it, helping students correctly relate the colours of the colour scale to temperature subsequently.

The FA sequence also extends students' learning. Representations serve as tools for constructing new meanings (Ainsworth, 2006). Comparing the colour scale and touch scale can help students understand objectivity and accuracy in measurement, but this potential requires teacher scaffolding (Tippett, 2016). In Excerpt 3, students preferred the colour scale for measuring temperature but struggled to articulate their reasoning, presenting an opportunity for the teacher to introduce the technical term "accuracy.

3.6.2 *Strategies for Engaging Students in the FA Sequence*

From the case example, we identify some strategies for conducting the FA sequence that are useful in a representation-constructing activity.

3.6.2.1 Dialogic Talk

The excerpts demonstrate some degree of dialogic talk particularly in the "explore" phase. Mortimer and Scott (2003) define dialogic talk to be an interaction whereby more than one idea is allowed into the conversation. In the case study, students were able to freely contribute their ideas without evaluation, particularly during the explore phase. Such a dialogic space is important for formative assessment whereby the first step is to establish where students are in their learning (Black & Wiliam, 2009). A dialogic space allows students to feel safe to contribute their ideas which in turn gives teachers a deeper insight into students' thinking.

3.6.2.2 Transduction and Transformation of Representations

The excerpts illustrate how concept learning is orchestrated through processes of transduction and transformation of representations, akin to how scientists develop new concepts (Gooding, 2004; Nersessian, 1992) and how high school students reason when constructing scientific explanations (Yeo & Gilbert, 2014, 2022). These processes bridge abstract scientific ideas with students' empirical experiences of the real world. In Excerpt 1, a transduction process unpacks the epistemic meaning of "hotness." By referencing the touch scale in Turns 11 and 21 which represents cups of water in iconic images and an arrow defined by "hotness" respectively, students are reminded of both the embodied activity and the abstract concept of ranking water's hotness.

Similarly, Excerpt 2 shows another transduction process used to address misconceptions, often arising from the abstraction of scientific ideas. Here, the teacher refers to the touch scale (Turn 75) and connects the physical activity with its conceptual meaning, emphasizing that the scale refers to temperature. By drawing parallels between the touch scale and the color scale, the teacher helps students recognise that both scales measure how hot something is.

These excerpts highlight that concept development involves processes of transduction and transformation across various modes to connect different meanings. This process can be challenging for students, making it essential for teachers to provide support in facilitating these transformations. The excerpts show how orchestrating these processes look like.

3.6.2.3 Direct Instruction

Student-centred representation constructing activity does not preclude direct instruction. We see instances of this in Excerpt 1 when the teacher showed the scale when asked what a scale is. In this case, the teacher must have found the visual representation of the scale to be apt in describing it. We also found another instance of direct instruction in Excerpt 3. This was the case when the students were not able to use the correct scientific terminology in describing their preference for using colour scale. As such, the teacher didactically introduced the term "accurate" to the students. As these scientific terminologies (e.g., scale and accurate) may not be familiar to the students, it will be difficult for students to arrive at these technical terms by themselves. Direct instruction may be necessary when it relates to the conceptual meanings of scientific conventions and terminologies.

3.7 Conclusion

In conclusion, the chapter introduces a formative assessment sequence as a strategy to support teachers in orchestrating talk when addressing students concept learning in a representation-construction activity. As students generate and work with representations, they interpret meanings of and for the representations based on their motivated purpose (Kress & van Leeuwen, 2006, which may not be aligned with the canonical ones or those intended by the curriculum. In a similar vein, the many facets of concepts and their representations may make it challenging for teachers to know what to do and where to start. The formative assessment sequence "explore—work-on—review" provides a structure to guide teachers in exploring students' thinking and reasoning, and a means for the teacher to align students' meanings with the canonical ones (Black & Wiliam, 2009. As shared by the teacher, "Before I come into contact in, … no guiding framework … to frame your thinking in a discussion. … I would say before that would be more haphazard and random. Now there's a guiding framework … you work on the key ideas, you err, explore then you work on the key ideas, then finally you review it before you move on to the next part. I find that quite useful."

Acknowledgements This study is funded by a research grant from the National Institute of Education
(Singapore) (AFR02/15JY).

References

Ainsworth, S. (2006). DeFT: A conceptual framework for considering learning with multiple representations. *Learning and Instruction, 16*(3), 183–198. https://doi.org/10.1016/j.learninstruc.2006.03.001

Ainsworth, S., Prain, V., & Tytler, R. (2011). Drawing to learn in science. *Science, 333*(6046), 1096–1097. https://doi.org/10.1126/science.1204153

Black, P. J., & Wiliam, D. (2009). Developing the theory of formative assessment. *Educational Assessment, Evaluation and Accountability, 21*(1), 5–31.

Bransford, J. D., Brown, A. L., & Cocking, R. R. (2000). *How people learn: Brain, mind, experience, and school.* National Academy Press.

Britsch, S. (2013). Visual language and science understanding: A brief tutorial for teachers. *Australian Journal of Language and Literacy, 35*, 17–27.

Driver, R., Guesne, E., & Tiberghien, A. (1985). Some features of children's ideas and their implications for teaching. *Children's Ideas in Science*, 193–201.

Duit, R., & Treagust, D. F. (1995). Students' conceptions and constructivist teaching approaches. In B. J. Fraser & H. J. Walberg (Eds.), *Improving science education* (pp. 46–69). The National Society for the Study of Education.

Eilam, B. (2012). *Teaching, learning, and visual literacy: The dual role of visual representation in the teaching profession.* Cambridge University Press.

Gooding, D. C. (2004). Envisioning explanations–the art in science. *Interdisciplinary Science Reviews, 29*(3), 278–294.

Gray, R., & Rogan-Klyve, A. (2018). Talking modelling: Examining secondary science teachers' modelling-related talk during a model-based inquiry unit. *International Journal of Science Education, 40*(11), 1345–1366. https://doi.org/10.1080/09500693.2018.1479547

Hamnell-Pamment, Y. (2024). The role of scientific language use and achievement level in student sensemaking. *International Journal of Science and Mathematics Education, 22*(4), 737–763.

Hubber, P., Tytler, R., & Haslam, F. (2010). Teaching and learning about force with a representational focus: Pedagogy and teacher change. *Research in Science Education, 40*(1), 5–28.

Khan, S. (2011). What's missing in model-based teaching. *Journal of Science Teacher Education, 22*(6), 535–560.

Kress, G. (2010) *Multimodality: A social semiotic approach to contemporary communication.* Routledge.

Kress, G., & van Leeuwen, T. (2006) *Reading images: the grammar of visual design* (2nd edn.). Routledge.

Latour, B. (1999). *Pandora's hope: Essays on the reality of science studies.* University Press.

Lee, T. D., & Jones, M. G. (2017). Elementary teachers' selection and use of visual models. *Journal of Science Education and Technology, 27*, 1–29.

Lemke, J. (1990). *Talking science: Language, learning, and values.* Ablex Publishing.

Mortimer, E., & Scott, P. (2003). *Meaning making in secondary science classrooms.* McGraw-Hill.

Nersessian, N. (1992). Constructing and instructing: The role of "abstraction techniques" in creating and learning physics. In R. Duschl & D. Hamilton (Eds.), *Cognitive psychology, and educational theory and practice* (pp. 48–68). State University of New York Press.

Paik, S. H., Cho, B. K., & Go, Y. M. (2007). Korean 4- to 11-year-old students conceptions of heat and temperature. *Journal of Research in Science Teaching, 44*(2), 284–302.

Pappas, C. C., & Varelas, M. (2009). Multimodal books in science-literacy units: Language and visual images for making meaning. *Language Arts, 86*, 201–211.

Posner, G., Strike, K., Hewson, P., & Gertzog, W. (1982). Accommodation of a scientific conception: Toward a theory of conceptual change. *Science Education, 66*, 211–227. https://doi.org/10.1002/sce.3730660207

Prain, V., & Tytler, R. (2013). Representing and learning in science. In R. Tytler, V. Prain, P. Hubber, & B. Waldrip (Eds.), *Constructing representations to learn in science* (pp. 1–14). Sense.

Tang, K.-S., & Tan, S.-C. (2017). Intertextuality and multimodal meanings in high school physics: Written and spoken language in computer-supported collaborative student discourse. *Classroom Discourse, 8*(1), 19–35. https://doi.org/10.1080/19463014.2016.1263576

Thomaz, M. F., Malaquias, I. M., Valente, M. C., & Antunes, M. J. (1995). An attempt to overcome alternative conceptions related to heat and temperature. *Physics Education, 30*(1), 19.

Tippett, C. D. (2016). What recent research on diagrams suggests about learning *with* rather than learning *from* visual representations in science. *International Journal of Science Education, 38*(5), 725–746. https://doi.org/10.1080/09500693.2016.1158435

Tytler, R., Hubber, P., & Prain, V. (2013). Structuring learning sequences. In R. Tytler, V. Prain, P. Hubber, & B. Waldrip (Eds.), *Constructing representations to learn in science* (pp. 51–65). Sense.

Tytler, R., Prain, V., Hubber, P., & Waldrip, B. (2013). Constructing representations to learn in science. *Sense.*

Vosniadou, S. (2007). Conceptual change and education. *Human Development, 50*, 47–54. https://doi.org/10.1159/000097684

Vygotsky, L. S. (1978). *Mind in society: The development of higher psychological processes.* Harvard University Press.

Waldrip, B., & Prain, V. (2013). Teachers' initial response to a representational focus. In R. Tytler, V. Prain, P. Hubber, & B. Waldrip (Eds.), *Constructing representations to learn in science* (pp. 15–30). Sense Publishing.

Yeo, J., Tan, K. C. D., & Tan, P. H. (2020). From images to writing: A formative assessment approach for developing understanding of abstract concepts in primary science. MOE Academies Fund: Closure Report. National Institute of Education.

Yeo, J., & Gilbert, J. K. (2014). Constructing a scientific explanation—A narrative account. *International Journal of Science Education, 36*(11), 1902–1935.

Yeo, J., & Gilbert, J. K. (2022). Producing scientific explanations in physics—A multimodal account. *Research in Science Education, 52*(3), 819–852. https://doi.org/10.1007/s11165-021-10039-1

Chapter 4
Spatial Reasoning in Explorative, Creative, and Playful Science and Design Learning

Remke Klapwijk and Caiwei Zhu

Abstract Spatial reasoning is often described as the capacity to understand, interpret, and manipulate visual and mental representations of objects in three-dimensional space and plays an important role in science and design activities. It serves as a cognitive cornerstone that underpins various aspects of learning, problem-solving, and creativity. In this chapter, we explore the nature of spatial reasoning and demonstrate how it can be nurtured through playful science, design, and making activities. First, we explain what spatial reasoning is and why it is important to pay explicit attention to fostering spatial reasoning skills development in curricula. Next, we describe four educational strategies: using models, diagrams, spatial language, and engaging in sketching and gesturing, all of which can potentially help students think more spatially in science and design lessons. We then focus on the relation between spatial reasoning and designing, illustrating that spatial reasoning is required at all stages of the design process, including empathizing, ideating, as well as in prototyping and testing. This is clarified through examples of playful design activities for preschoolers and primary school students. Spatial reasoning is often implicitly present in science and design education. However, when educators appropriately scaffold learning processes that require spatial reasoning, learners are more likely to develop their spatial reasoning skills.

4.1 Introduction

Spatial reasoning[1] is ubiquitous in everyday life. Finding our way in a shopping mall, estimating how much paper we need to wrap a present, making a jigsaw puzzle, and reading instructions to assemble flat-pack furniture are a few examples of how

R. Klapwijk (✉) · C. Zhu
Delft University of Technology, Delft, Netherlands
e-mail: r.m.klapwijk@tudelft.nl

C. Zhu
e-mail: c.zhu-1@tudelft.nl

[1] In this chapter, we use the terms spatial reasoning and spatial thinking interchangeably.

S. Earle et al. (eds.), *Primary Science Learning for Children, Teachers, and Communities*, https://doi.org/10.1007/978-981-95-0561-6_4

we engage with the world in spatial ways. In many professions, such as artists, scientists, designers, engineers, technicians, and mathematicians, strong spatial skills are needed. It is worthwhile to invest in the development of spatial reasoning skills in science, technology, engineering, and mathematics (STEM) education as these skills can be developed through learning and practising (Uttal et al., 2013a).

There are a variety of ways to support students' spatial reasoning skills development in science education (Newcombe, 2017). For instance, when teachers use models and gestures in the classroom, such as modeling the solar system and gesturing about how the Earth rotates, students can visualize and imagine the scientific processes in a more concrete and tangible way. Sensing exercises followed by probing questions, for example, encourage students to become better explorers of the world. Students might notice that the beak of a small tree finch is long and skinny with a bent beak that looks like a needle-nose plier, while the big tree finch has a strong, wide beak that resembles the beak of a parrot (Poelman, 2015). Probing questions, such as 'which of these birds eats insects and which uses its beak to crack nuts', can help students relate forms seen in nature to their functions.

Design and technology education also benefits from spatial scaffolding. In any design process, students first explore the situation to understand the design problems, then think divergently to generate multiple, varied solutions and select the most promising ones to develop further (Cross, 2006). Ultimately, they create solutions that are original and valuable to the students themselves (Craft, 2003). Throughout these processes, imagining solutions in one's mind and elaborating on them through sketches and rapid prototypes engage novice designers' spatial thinking skills (Zhu et al., 2024). If the design does not work as intended, children are still learning, especially when they are given room and support to improve their design ideas (Sonneveld et al., 2024).

In this chapter, we will explain how many of the science and design activities are spatial in nature and how educators can apply age-appropriate approaches to integrate spatial reasoning in science and design education, from preschool through the end of primary schools.

4.2 What is Spatial Reasoning?

Spatial reasoning involves understanding how objects, including ourselves, relate to one another and to the spatial environment (Gifford et al., 2022). It includes interpreting, creating, and manipulating visual-spatial representations both mentally and through various medium. Here are some examples of circumstances where we need spatial reasoning in our daily lives:

- When you are figuring out how to reach a certain store in a shopping mall, you are thinking about positions and paths.
- When you are designing a pop-up card, you may mentally visualize how the pop-up card will look after cutting the figures.

– When you see a graph of the cross-section of a tree trunk, you can use this to imagine how the inside of the tree looks.

When we talk about mathematics or language literacy, it is clear to most of us that there are many different elements involved. The same is true for spatial reasoning or spatial thinking, as it consists of many elements (Buckley et al., 2018). Cognitive scientists make a distinction between large-scale and small-scale spatial reasoning (Newcombe & Shipley, 2014; Uttal et al., 2013a). Large-scale abilities relate to orienting oneself in a larger space, such as way-finding, map reading, and creating mental maps of routes. Small-scale abilities involve thinking about transformations within objects or structures, like positioning Lego blocks to build a structure or determining how many students should sit on one end of a seesaw when a parent sits on the other end.

Spatial reasoning skills undergo considerable development during infancy and early childhood (Newcombe & Frick, 2010) and develop throughout the entire life (Uttal et al., 2013b). Playing with spatial toys has been part of education since Friedrich Fröbel (1782–1852) developed hands-on, playful educational materials featuring mechanical spatial elements, such as construction sets of wooden blocks (Liebschner, 2006). Meanwhile, the school and its surroundings can be considered as a spatial environment that students use to develop observation, sensory, tactile, and perspective-taking skills. In addition, physical play and exercises help children develop the ability to sense and perceive the movement and position of their bodies, developing spatial-motor skills (Maresch & Sorby, 2021).

Recently, Ramey and Uttal expanded the definition of spatial reasoning from focusing solely on forming and manipulating images in our minds to including both internal cognitive processes and external objects or spatial representations, such as prototypes, models and diagrams (2017, p. 278). This broader definition is especially helpful when we discuss science and design education, where students actively engage with the material world, do experiments, make sketches and prototypes, and continuously iterate between visualizing and refining their design ideas—both mentally and physically—through discussion, sketching, and prototyping. A recent meta-analysis found that the use of tangible, concrete materials during spatial training sessions appear to have a positive effect on students' math performance later on (Hawes et al., 2022). Therefore, design and science activities, which are often fundamentally material in nature, offer a promising route to develop spatial reasoning skills.

In line with the growing emphasis on how mind and body work together in developing knowledge, skills, and understanding (Barsalou, 2008), the Assessment of Performance Unit (APU) model of iterative design processes (Fig. 4.1) highlights the alternation between internal and external cognitive processes. Internal processes like speculating and reflecting, represented on the left side of the model, involve using the mind's eye to imagine solutions. External processes happening outside of the head, shown on the right, include peer discussions, gestures, and tangible representations such as sketches and prototypes (Ramey & Uttal, 2017; Van Mechelen,

2016). Through these iterative internal and external cognitive processes, hazy ideas about design or STEM problems become increasingly tangible.

Both science and design activities are multi-modal in practice: students use language, gestures, sketches, as well as tangible and digital materials, including two-dimensional and three-dimensional models, to explain their ideas to peers, teachers, and themselves. Much research on spatial skills training in STEM disciplines has focused on solving problems that have only one correct answer (Zhu & Klapwijk, 2024), such as determining how a certain structure appears after being rotated by 90 degrees in a multiple-choice question. In design and technology projects, however, students learn to develop new or original solutions through divergent and convergent thinking, while using spatial thinking in creative ways (e.g. Sonneveld et al., 2024; Zhu et al., 2024). Therefore, integrating spatial thinking in science and design classrooms provides teachers with opportunities to teach both analytical and creative

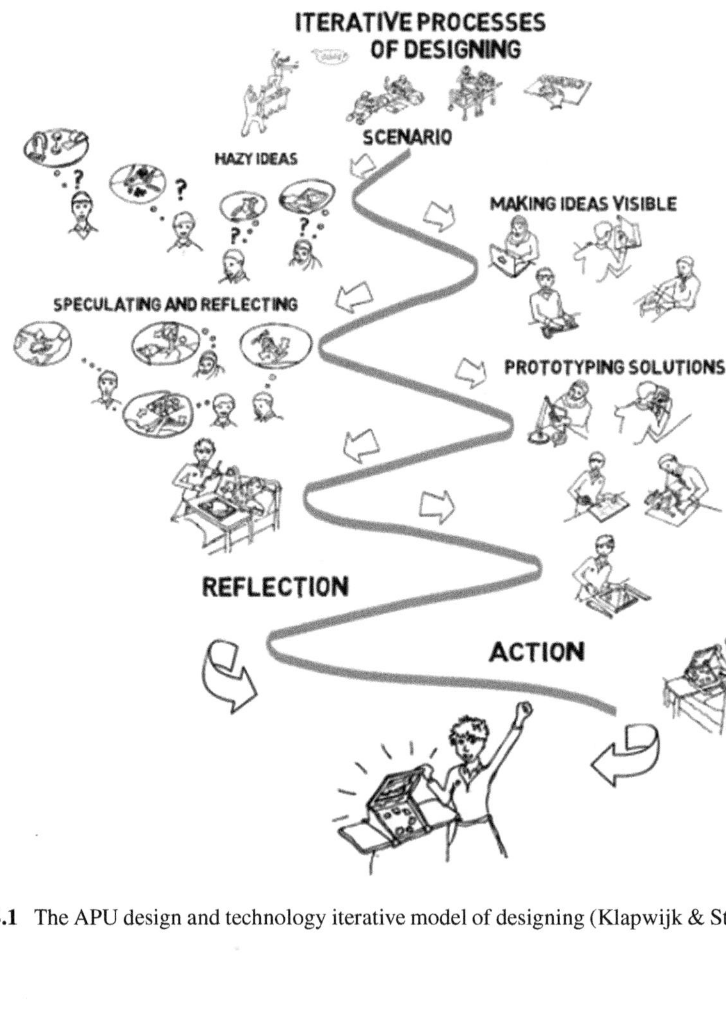

Fig. 4.1 The APU design and technology iterative model of designing (Klapwijk & Stables, 2023)

problem-solving, which will be elaborated on in the following sections on educational strategies and stories of practices.

4.3 Why Spatial Reasoning is Important in the Classroom

Spatial thinking plays an important role in scientific breakthroughs and technological advancement (Kell et al., 2013). The U.S. National Research Council described the process of spatial thinking as 'using the properties of space as a vehicle for structuring problems, for finding answers, and for expressing solutions' (2006, p. 3). Such a thinking process resembles what is needed to conduct scientific inquiry, as well as design and engineering problem-solving.

STEM learning and spatial reasoning skills development go hand in hand, with each positively influencing the other (Buckley et al., 2018; Wai et al., 2009; Zhu et al., 2023a). Prior research indicated that primary school students who have higher spatial ability tend to perform better in mathematics (Gilligan et al., 2019) and science (Hodgkiss et al., 2018). For students who are new to science subjects, such as geology, spatial ability is one of the important predictors of the learning outcome (Hambrick et al., 2012). Compared to students with higher spatial ability, those with lower spatial ability may face more challenges visualizing and comprehending spatially challenging concepts in STEM, potentially leading to poorer performance (Uttal & Cohen, 2012). Altogether, it appears that having a higher level of spatial ability is generally helpful for STEM learning, and spatial ability may even serve a role as important as the 'gatekeeper' in early STEM learning (Uttal & Cohen, 2012; Uttal et al., 2013b). In the long run, students with high spatial ability are more likely to pursue STEM subjects and careers (Wai et al., 2009) and perform better in engineering classes (Sorby, 2009).

For these reasons, it seems valuable to pay attention to spatial reasoning in science and design classrooms and to engage students in spatially challenging activities. Luckily, spatial ability can be developed through appropriate activities and opportunities (Uttal et al., 2013a) that can be integrated into pre-school and elementary classrooms as well as the STEM curriculum (Lowrie et al., 2018).

Some research has revealed gender differences in spatial reasoning skills, showing that boys outperform girls in certain tasks, such as mental rotation tasks (Neuburger et al., 2015). However, in a recent meta-analysis, Lauer et al. (2019) found no significant gender difference in mental-rotation skills among preschoolers, though a small male advantage emerged in students between the ages of six and eight. This suggests that, while these results do not exclude the possibility that biological influences may contribute to the gender gap, other factors such as gender roles and social norms (Bartlett & Camba, 2023), as well as test types and formats (Neuburger et al., 2015) likely play a role in gender differences in performance.

For a teacher, it is important to challenge all students with spatialized activities at the appropriate level and realize that certain girls will be top performers while some boys may struggle with spatial reasoning. Through our collection of stories of practice, we will highlight several good practices that exercise and cultivate spatial reasoning skills in science and design classrooms.

4.4 Four Educational Strategies to Involve Spatial Reasoning in Science and Design Classrooms

Educators, often unknowingly, provide learners with a variety of spatialized science and design activities while also utilizing a variety of spatial learning tools in science classrooms, including spatial language, diagrams, graphs, models, analogies, as well as action-based tools like gesturing and sketching (Newcombe, 2017). In this section, we explicate these tools so you can use them more frequently in your classrooms.

Strategy 1. Experimenting with manipulatives and models

Connecting learning with doing in science classrooms often stimulates students' interest and curiosity, keeps them motivated and engaged, and connects their knowledge with practices (National Research Council, 2012). By using tangible materials to experiment with, students can make abstract and symbolic models in science classrooms, such as models of sound waves or electrical circuits. These models act as tools "for thinking with, making predictions, and making sense of experiences" (National Research Council, 2012, p. 56). They further allow students to connect the observed with the unknown to test out their hypotheses through hands-on trials, formulate explanations, and grasp abstract science concepts. Sixth-graders who experimented with building different robotic models in an educational robotics project showed improvement in their spatial orientation task performance and exercised their mechanical reasoning skills (Julià & Antolì, 2018).

A simple yet fun example would be modeling sound waves with slinkies (e.g., see resources available on Exploratorium.edu (2024), 'Slinky in Hand: Physics & Waves Science Activity'). Through compressing, stretching, and creating waves with slinkies, students can learn about how sound travels and the different properties of sound waves. Meanwhile, to connect the movement of the slinky with the sound wave to understand abstract concepts such as antinodes of slope or frequency of oscillation, students also need spatial thinking to visualize the effects of different motions, and how such effects can relate to and be represented through 2D symbolic representations of sound waves in their textbooks.

Strategy 2. Using diagrams and graphs

Learning to understand, use, and create scientific diagrams is an important skill to be cultivated through the science curriculum in primary education, especially in the upper primary years. Diagrams are important scientific reasoning tools that contribute to science learning outcomes and engage students' spatial reasoning skills (Cromley et al., 2013). Whether it is learning about living things and their habitats, the solar system, or learning about physics topics such as sound, light, forces, and electricity, diagrams are not only visually exciting and challenging for students but can also inspire engaging classroom activities. For instance, the diagram (Fig. 4.2) below is a learning activity from the Oxford Primary Illustrated Science Dictionary Free Resources (2019) that probes students' understanding of the food web.

Having knowledge about the different organisms mentioned in this activity is perhaps not enough for students to successfully fill out the blanks. Students need to realize how information about predator–prey relations is visualized in the diagram. The arrows in the diagram point from one blank to another to indicate either unique or shared predator–prey relationships between these organisms. Filling out one blank with an organism's name would require students to think about how this blank is positioned

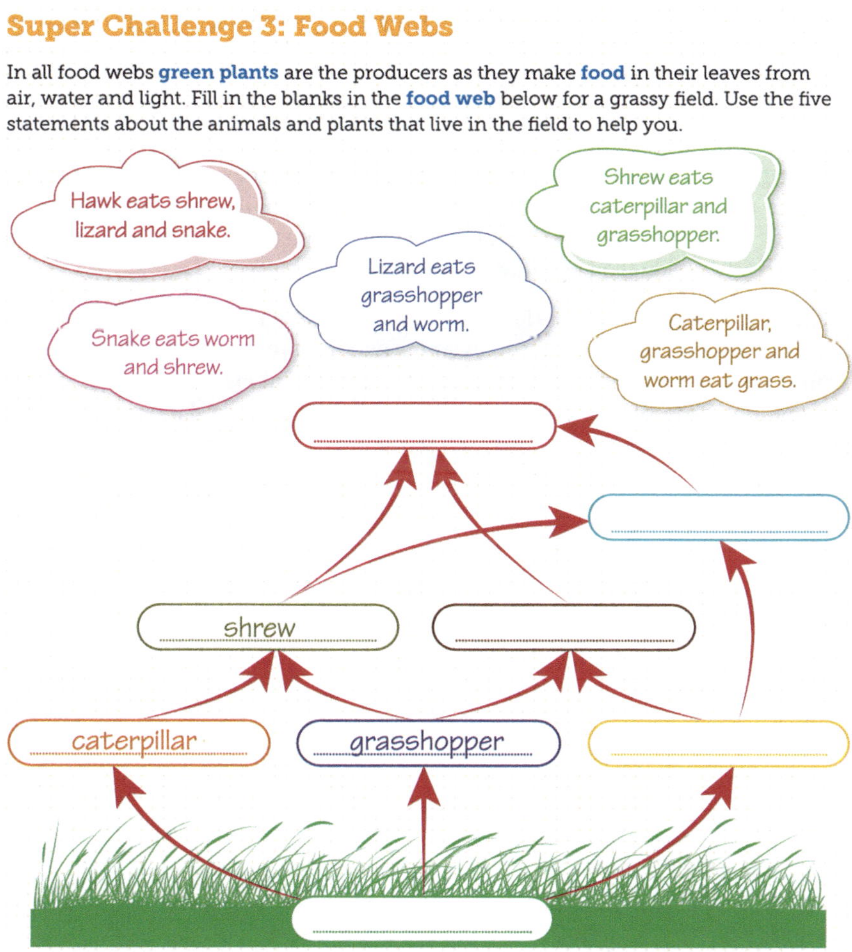

Fig. 4.2 Learning about and visualizing the food web through a diagram (Oxford Primary Illustrated Science Dictionary Free Resources, 2019)

regarding the entire diagram and regarding other blanks. Therefore, in this diagram learning process, students need to constantly think about how the relationships among the organisms can be visualized on this food web and the spatial relations among the blanks.

In science classes, educators can use drawings to formatively assess their students' understanding of scientific concepts. For example, at the start of a new subject or during the activities, students can create drawings of how they understand a scientific concept, such as how the heart and the circulatory system work, to see what students already know and if there are any pre- or misconceptions prevalent.

Ten to fifteen-year-old students were asked to draw a heart and to describe its parts and function in the human body (Fančovičová & Prokop, 2019). Students' drawings of the heart were in three shapes, some students drew it as Valentine heart, others drew it as a circle and other students drew the heart in its real shape (Fig. 4.3). Educators can evaluate these drawings by asking questions such as: Do students divide the heart and display specific parts of the heart? Are there signs of veins and arteries, like outlets from and to the heart? Do they schematically divide the heart into upper left and right atria and lower left and right ventricles? Do they realize that the heart is connected to the respiratory system? In practice, each pupil could individually draw a scientific process, then come together in small groups to share their drawings and collaboratively create the best drawing version.

In summary, diagrams seen in learning materials as well as making diagrams can potentially be good opportunities to examine and nurture students' understanding of scientific concepts while exercising students' spatial thinking. This also accounts for charts, graphs and other visuals that represent information or data in visual-spatial ways.

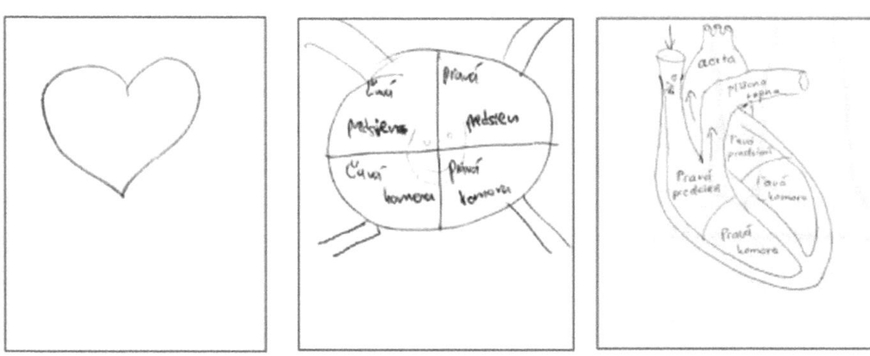

Fig. 4.3 The shapes of heart in students' drawings (*Source* Fig. 4.1 in Fančovičová & Prokop, 2019, p. 553)

Strategy 3. Using spatial language

Educators incorporate spatial language and gestures in many of their teaching activities. For example, when teaching writing, a teacher might say and gesture, 'Around, up, down, and tick.' Spatial language serves as a cognitive tool that facilitates the representation and reasoning about spatial information (Gentner, 2016). Exposure to spatial language enhances students' understanding of spatial relations, shapes, forms and transformations. Newcombe (2017) wrote, '*language can highlight dimensions that might otherwise be less likely to be noticed and remembered. For instance, students find it easier to remember left–right relation when someone has drawn attention to the relationship using language, even when the language is not strictly relevant to the precise spatial arrangement, for example when one side is simply said to be prettier than the other (p. 25).*'

To cultivate a habit of using more precise spatial vocabulary in the classroom, teachers could consider what they want to communicate and aim to be specific about the shapes and spatial relations of objects. For example, you might say, 'Could you hand me the rectangular book?' or 'Look at the pretty red block that is above the blue one.' Using the following five categories of language will help teachers incorporate more spatial language to their talk in science or classes (Gamarra & Nguyen, 2024). As a result, educators will draw attention to the spatial aspects of objects and spatial relations between them.

- Description of shapes (e.g., square, circle, cube, prism, wobbly).
- Comparison (e.g., bigger, smaller).
- Spatial prepositions (e.g., between, on, in, under).
- Movement (e.g., "the electrons flow", "when one gear spins, the other will spin too").
- Location (e.g., "on the first page", "perpendicular to").

Strategy 4. Encouraging sketching and gesturing

Two additional action-based spatial reasoning tools are sketching and gesturing. We tend to think of these two as tools to communicate or represent thoughts and solutions. Yet besides this, they are also reasoning tools that offer visual and motor input into the thinking process, influencing and supporting learning (Newcombe, 2017). Figure 4.4 shows an example of two 11-year-olds' notes with sketches when recording their observations of nature and thinking about what they could design based on forms and functions seen in nature. From observing the shape and learning about the functions of tree roots, these students came up with a claw-like tent hook design that can help hold tents firmly in the ground like tree roots do for trees, while also being able to 'contract for space efficiency'. Sketching while thinking ensured that the students were actively observing, documenting, analyzing, interpreting, and organizing knowledge about the visual-spatial properties seen in nature, meanwhile sparking further inquiries.

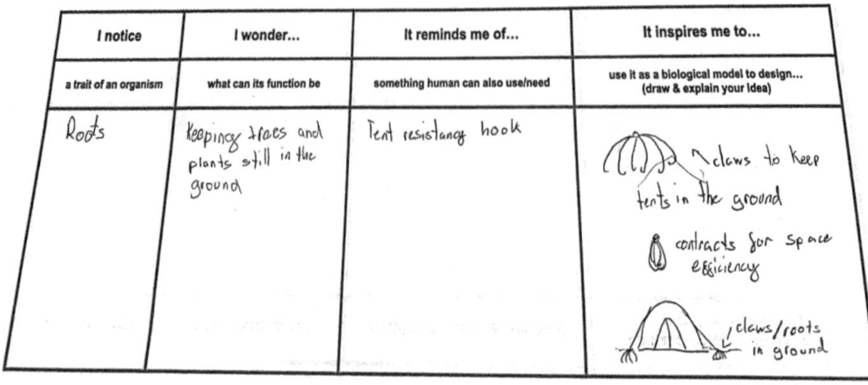

Fig. 4.4 Children's observations of nature in a biomimicry design project (*Source* Fig. 4.3 in Zhu et al., 2024)

Besides sketching, both cognitive and design researchers have emphasized the importance of gesturing in learning to think spatially and visualizing ideas. Novice students who were encouraged to gesture while thinking through geology problems, compared to those who were not allowed to gesture, showed improved performance in reasoning about 3D diagrams and spatial relationships (Atit et al., 2015). Meanwhile, gestures, along with conversations, drawings and models, constitute the language of design (Cross, 2006), and are often used by designers when discussing visual representations, such as models, sketches or drawings (Jornet & Roth, 2018). In design and science education, gestures can be used to explain processes (e.g., gesturing the opening and closing of a pine cone's scales) or to describe how an envisioned product will look like (e.g., gesturing that an additional piece of fabric with folds will be used to lengthen a skirt). Even students new to design have been observed using gestures to talk about their sketches and prototypes, explaining to peers and teachers what they envisioned in their designs (Zhu et al., 2024).

An additional benefit of sketching and gesturing is that they can help reduce the cognitive workload required to process complex spatial information. For instance, when sketching arrows to indicate the direction of gravity and static friction to solve a physics problem (Hubber et al., 2010) or gesturing to mimic the movement of tectonic plates in geology (Atit et al., 2014), students no longer need to constantly keep these pieces of information in their minds, and can direct their freed-up cognitive resources to solving problems and generating new lines of inquiry.

In summary, spatial reasoning can be supported and developed in science and design classrooms through:

1) Experimenting with manipulatives and models
2) Using diagrams and graphs
3) Using spatial language
4) Encouraging sketching and gesturing

Let us now look into a number of spatial design activities in which learners use spatial reasoning in a creative way.

4.5 Explorative, Creative, and Playful Design Activities

As outlined in the framework for K-12 science and engineering education developed by the U.S. National Research Council (2012), using design practices to seek solutions for real-life engineering and technology problems and to meet the needs of individuals and society can foster meaningful learning and practice of scientific knowledge. In design, several things are considered important (Klapwijk & Stables, 2023; Lawson & Dorst, 2009). First, the design process progresses through a sequence of phases: it typically begins with problem exploration, followed by the formulation of goals and design criteria, after which one or more solutions are constructed. Second, progression occurs in iterations: several steps are repeated as insights grow. For example, after developing an initial series of solutions, a new design criterion may be discovered, or it may become clear that the problem is not yet well understood, indicating that further investigations are needed. Through research we conducted as part of the SellSTEM (Spatially Enhanced Learning Linked to STEM) project (www.sellstem.eu), we noticed several benefits brought by integrating design activities into STEM classrooms.

4.5.1 Story-Based Design Assignments in Pre-School

In pre-school, play and storytelling are important means of learning. Sonneveld et al. (2024) developed a design task for pre-schoolers around a dog named Boris using a picture book. Children were asked to create a new home for Boris, because the father of the family was allergic to dogs. This project showed that spatial reasoning skills are needed at many stages of the design process, during problem exploration (Boris needs a roof and walls to protect him), while building design prototypes, and during prototype testing. Interestingly, this research revealed that the problems preschoolers encountered while designing and making were related to their play preference. Many of the preschoolers with a preference for role play constructed a building around Boris. They discovered in the testing stage that Boris could not leave the building anymore and was locked in. Often these preschoolers also noticed that the dog was sitting in the dark because they had not thought about adding a door or windows to their structure. Children who had a preference for construction play encountered another spatial problem that their dog was usually outside their built structure and could not enter.

There are many other storybooks available that contain spatial design problems. A recent study by Mishra and Klapwijk (2024) described how preschool educators developed design assignments from storybooks by identifying key characters in storybooks and the problems these characters encountered. Next, the teachers applied a lesson study approach in which one teacher gave a spatialized design lesson and another teacher observed and were interviewed post-lessons. Through observations and interviews, educators better understood the level of their students' creative spatial thinking and the types of scaffolding (including the four strategies described in Sect. 4.4) they could apply to support students' spatial skills development. This research showed that educators were often surprised by their students—those whom they thought would have low spatial ability performed well, while students expected to be high performers had difficulties in generating and elaborating spatial design ideas (Mishra & Klapwijk, 2024). Although we cannot generalize from this study, the results indicated that teachers are not always aware of the spatial ability of their students and that focused observation and reflection help them better understand their students.

4.5.2 Real-Life Design Assignments

It is also possible to use design problems related to the everyday experiences of learners. In preschool classes, drinking cups often topple over. Why not ask preschoolers to develop a solution using Knex, building blocks and crafting materials? We did this in a preschool in Rotterdam. We noticed that although the girls would not play with the available Lego during free play, they were very engaged in our design activity and came up with creative solutions. During understanding the problem, generating ideas, making the design, and testing, spatial reasoning was applied. Some students built a wall to prevent the cup from falling; however, none of the students in our study realized that a cup can fall in all directions. The teacher and researcher asked them what would happen if the cup would fall in another direction (see Fig. 4.5, Sonneveld et al., 2023). Even after the demonstration, none of the students made a cup holder that would surround the cup completely.

4.5.3 Biomedical Design Challenges

Biomedical design challenges are spatial in nature as the students have to think about, for example, how they can help people with rheumatism and other health issues who struggle with everyday tasks like peeling carrots. We will demonstrate how all design processes in biomedical design challenges—ranging from problem exploring, generating ideas, building, testing, and improving prototypes (3D representations)—involve spatial reasoning.

Exploring and formulating

A design process can start with a problem or a wish, thinking about the user and the purpose. In this case, bus driver John loves to cook food, but it is challenging for

Fig. 4.5 Preschoolers thinking spatially to solve a design problem (Sonneveld et al., 2023)

him to do so with one arm. Through simulations followed by interviews, students experienced the problem. They needed to reason spatially to understand why peeling carrots and other tasks could be difficult when one has to do it with one arm only (Fig. 4.6).

Generating and selecting ideas

Creative people spend a lot of time in the early stages of a process. Therefore, a brainstorming session to think of as many solutions as possible is often organized. Figure 4.7 showcases numerous design ideas to enable a child who needs a wheelchair to play marbles, such as lowering the wheelchair or playing marbles at a table. These solutions were spatial in nature and used different spatial compositions to bring the marble and the flat space on which the marble will roll together. Additionally, a variety of spatial solutions were generated for moving the marble.

Elaborating and testing prototypes

Using sketches, a 3D prototype of a wheelchair that can be used to play marbles was built. During testing, the students tried to get a marble through the bent pipe. When the marble got stuck, the students thought that their design was not working. However, when the teacher asked the students if they were able to sit in the wheelchair, the students realized they had built a scale model of the wheelchair and used a small bead to test the piping system. In design, thinking in scale models is an important ability.

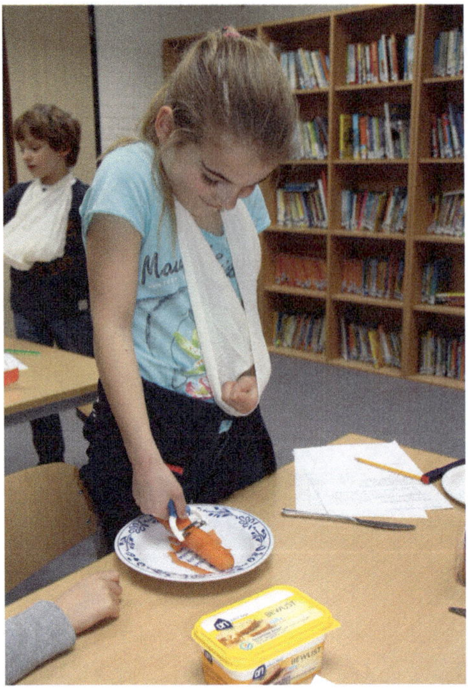

Fig. 4.6 Students simulating to empathize with the users

Providing students with scale models of people is one way educators can support scaling activities (Strand & Lutnæs, 2022).

In each design stage, spatial reasoning was engaged. Simulations, interviews with users, brainstorming sessions, sketching, prototyping, and testing all involve spatial reasoning. Educators can assess students' spatial reasoning through their discussions, representations, and gestures and notice where spatial reasoning is needed—like the teacher in our example who observed that the students did not scale correctly (Fig. 4.8).

4.5.4 Data Physicalization

In a world where data is ubiquitous, we encounter data every day in a variety of formats: numbers, tables, graphs, charts, and all kinds of 2D or 3D visualizations. Knowing how to read, understand, and interpret common data graphs such as bar graphs and pie charts are among the key learning goals in the primary science curriculum.

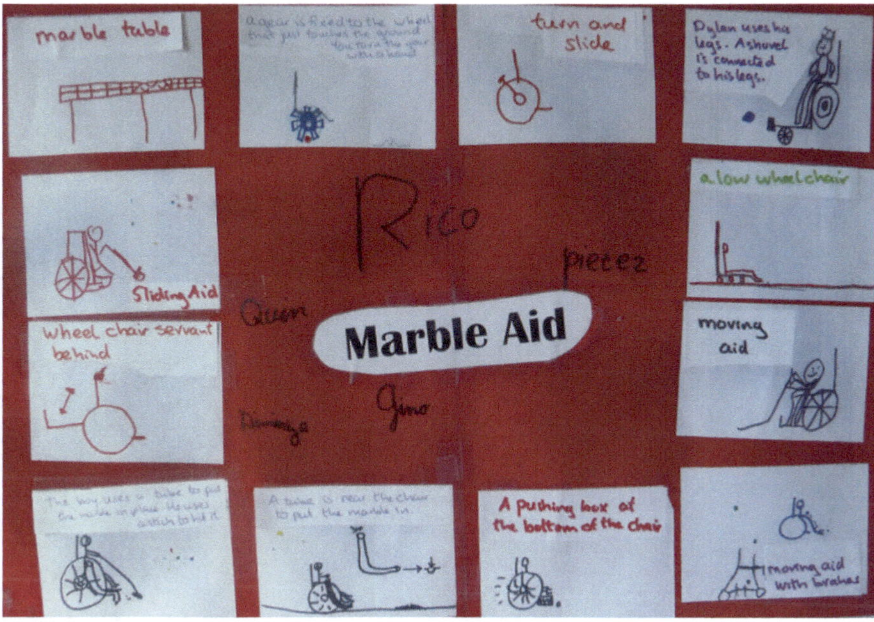

Fig. 4.7 Generating numerous solutions through brainstorming

Data physicalization means physical artifacts with geometrical or material properties that represent data (Jansen et al., 2015). By situating tangible materials in the 3D space, data physicalization models are not only innovative and engaging ways to communicate data obtained from a range of science subjects but also useful mediums to examine the patterns in and meaning of the data. Meanwhile, it engages spatial reasoning, as one would need to visualize non-tangible numbers as having tangible properties, such as length, depth, volume, and location, as well as to construct and communicate the visual-spatial relations between numerical values (Zhu et al., 2023b). Data physicalization appears to be an especially suitable classroom activity not only because of its engaging and versatile nature but also since analyzing and interpreting data as well as developing and using models are both listed as essential practices to be harnessed in K-12 science classrooms (National Research Council, 2012). Learning from and with data has the potential to immerse students in inquiries while the making of models enables students to test their hypotheses and make sense of learning through embodied experiences.

In our research project, we organized a data physicalization workshop with 11 and 12-year-olds (Zhu et al., 2023b). The workshop, featuring the Circular Economy theme to align with the ongoing module at the studied school, consisted of two 45-minute class sessions. The students were first introduced to data visualization and physicalization, read and selected data tables on waste production, and then spent around 40 minutes making their data physicalization model, followed by presenting their models in class for 15 min. They were encouraged to use a range of crafting

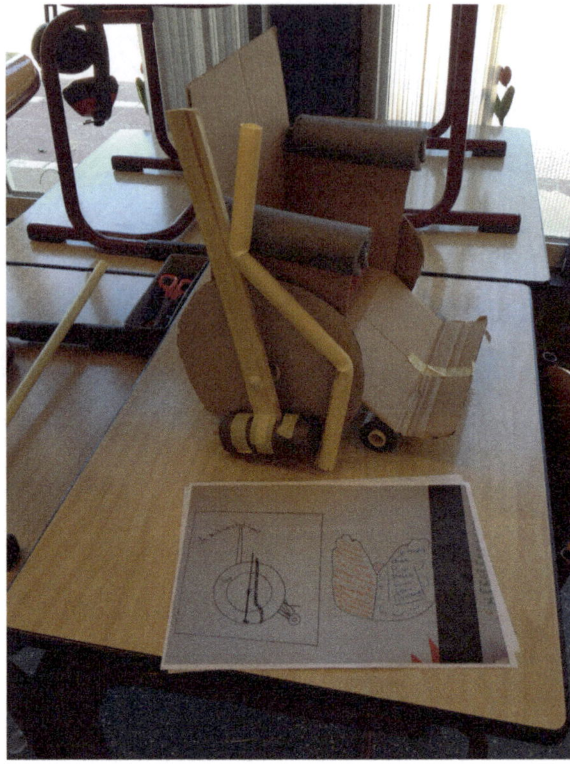

Fig. 4.8 From sketching to 3D prototyping

materials that are commonly seen in primary classrooms, including playdough, elastic bands, straws, pipe cleaners, paper cups and plates, yarn, beads, and balloons. This allowed students to not only playfully explore the different material properties but also to mold, shape, tie, stuff, or stack to arrange the materials in creative and meaningful ways. Below is an example showing how two students were actively exploring and experimenting when making their data physicalization model (Fig. 4.9). The conversation between these students proceeded as follows:

> Lia: I have an idea, so for compost and food we can fill it up until halfway or something (putting the pom poms in the cup to explain the idea) say this is the recycling thing. And fill it up that much. Like, really little, for 5%. Then, say, for this one (compost), we fill it almost halfway.
>
> Ada: So then, Um, How Much We Fill the Plastic?
>
> Lia: We Have to Fill Exactly the Same as the Paper.
>
> Ada: What if, to Make, Say, the Plastic Higher, We Can Put This on Top Here, and then We Put the Lid… and then We Can Fill
>
> That. It's Full of Plastic. So Wait, but is This 14% About?
>
> Lia: This One Has to Be Filled Slightly more…and the Plastic It Has to Be Filled Exactly the Same.
>
> Ada: That Looks Kind of the Same.

Lia: Yeah, that's Fine, that's Fine. Although, This Has to Be More Than This… the Tin is 5%…

Ada: So, This is 10% Compared to 5%. that's About Half.

Lia: Yeah, I think. Look, I had this fun experiment. Let us dump them all into one and see if that makes exactly a full cup. A full hundred percent. I do not think it will because I think it's a bit too much. That makes about a full 100%.

Ada: It's a Bit Over, Right?

Lia: Okay, I'm Taking Out One of Each just to Make It Slightly Smaller.

These two students chose to represent data tables of the percentages (by weight) of five different types of waste produced at their school in 2020. They explored the available materials and formed the idea of filling five paper cups with different amounts of different materials to symbolize the different types of waste produced at school. During the making, they frequently used language that demonstrated mathematical and spatial reasoning and continued to gauge, compare, and experiment with the amounts they added into each cup to reach a more accurate representation of the exact percentages of the data values and the ratio differences among those numbers. Through trial and error, they tested their understanding of numerical data

Fig. 4.9 Two students experimenting with materials to improve their data physicalization model (Zhu et al., 2023b)

values using objects that have visual-spatial properties, such as height, volume, and position.

Moreover, these students initiated an especially interesting experiment. They decided to dump all the materials they added to the five cups into one cup, to examine if these materials would fill up one paper cup that indicates 100%. This was a clever way of testing the hypothesis they had in mind as well as their mental model—that the estimated amounts of materials used to fill the five cups should be rather accurate and should be able to be combined together to represent a whole entity. Using tangible materials, they could visually compare their predictions about the physical embodiment of numerical numbers, analyze how accurate they were when mapping percentages onto volumes, and practice using mathematical reasoning.

After making data physicalization models, a majority of the students reported that they could comfortably interpret the meaning of numerical data, confidently communicate the meaning behind the numbers using their data physicalization models, and gained a better understanding of the different possible ways of representing data.

4.6 Discussions, Recommendations, and Opportunities for Practice

Most of the research on the development of spatial reasoning skills of primary school students has focused on analytical activities in the domain of mathematics education under the label of direct spatial training. In this chapter, we have demonstrated that science and design education are inherently spatial in nature and can be effectively used to develop spatial reasoning skills. These activities evoke spatial thinking in a rich and engaging context, allowing students to apply spatial reasoning skills during analytical and creative problem-solving activities in science and design. Compared to direct spatial training, integrating spatial thinking in science and design activities is a more authentic way to learn to think spatially and is more related to actual STEM practices.

We conjecture that the relation between science, design, and spatial reasoning is reciprocal. Science and design education contribute to the development of spatial reasoning skills, while the development of spatial skills will also contribute to the students' science and design capabilities. When educators pay explicit attention to the spatial aspects of science and design learning, we expect that students will develop their understanding of these subjects more effectively.

Although science and design are spatial in nature, educators can support the development of spatial reasoning skills in various ways. First, using age-appropriate science and design projects centered around real-life problem-solving will provide students with an authentic context to practice applying spatial skills to solve problems that have more than one correct answer. Moreover, appropriate use of models and manipulatives, spatial language, diagrams and graphs, as well as gestures and sketches will help educators to draw students' attention to the spatial aspects of the science and design activities they are engaged in, see Fig. 4.10 for an overview of the key strategies.

More research on the development of spatial reasoning through science and design is needed. Although there is substantial evidence that tools like spatial language, gestures, 2D- and 3D-representations are supportive, research on these tools in the context of primary science and design education is scarce. In addition, as these tools reflect the spatial reasoning processes of students, the ways in which students externalize their thoughts through language, gestures, or models may be used to formatively and summatively assess students' spatial understanding, benefiting future educational practices. Although (formative) assessment of spatial reasoning in science and design classrooms is not yet well-developed, educators can follow the spatial reasoning processes of their students by observing their interactions with the materials and their group discussions. Using prompts that help students evaluate the spatial dimensions of their work is recommended.

Lastly, spatializing the curriculum is not a difficult endeavor that requires extensive resources, as demonstrated by the data physicalization, biomedical, story-based, and real-life design projects. Resources for these projects are freely available on the Delft University of Technology website (https://www.tudelft.nl/en/wetenschapsknooppunt/your-turn).

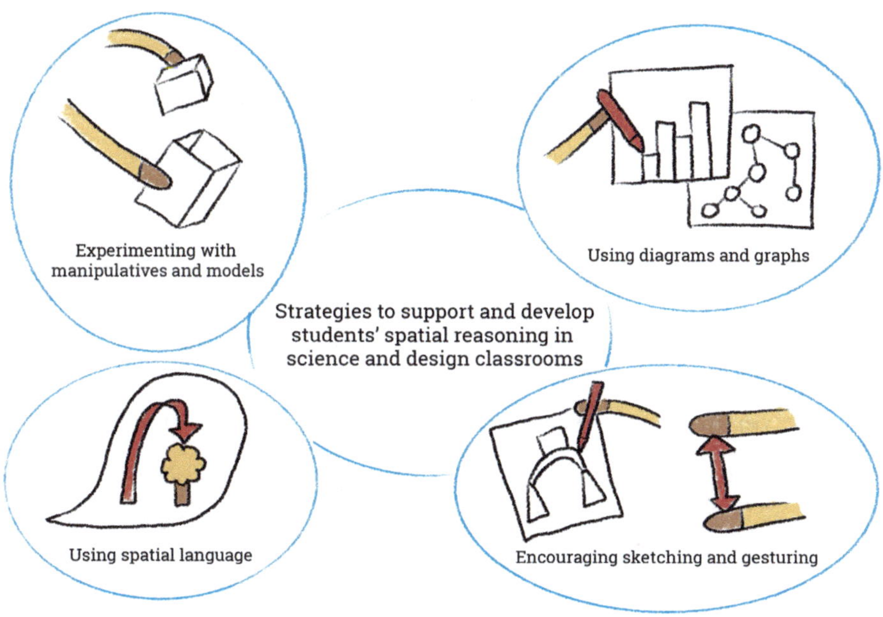

Fig. 4.10 Strategies to support spatial reasoning in science and design classrooms

References

Atit, K., Gagnier, K., & Shipley, T. F. (2015). Student gestures aid penetrative thinking. *Journal of Geoscience Education, 63*(1), 66–72. https://doi.org/10.5408/14-008.1

Atit, K., Shipley, T. F., & Tikoff, B. (2014). What do a geologist's hands tell you? A framework for classifying spatial gestures in science education.

Barsalou, L. W. (2008). Grounded cognition. *Annual Review of Psychology, 59*(1), 617–645.

Bartlett, K. A., & Camba, J. D. (2023). Gender differences in spatial ability: A critical review. *Educational Psychology Review, 35*(1), 8.

Buckley, J., Seery, N., & Canty, D. (2018). A heuristic framework of spatial ability: A review and synthesis of spatial factor literature to support its translation into STEM education. *Educational Psychology Review, 30*(3), 947–972.

Craft, A. (2003). Creative thinking in the early years of education. *Early Years, 23*(2), 143–154.

Cromley, J. G., Bergey, B. W., Fitzhugh, S., Newcombe, N. S., Wills, T. W., Shipley, T. F., & Tanaka, J. C. (2013). Effects of three diagram instruction methods on transfer of diagram comprehension skills: The critical role of inference while learning. *Learning and Instruction, 26*, 45–58.

Cross, N. (2006). *Designerly ways of knowing*. Springer.

Exploratorium.edu. (2024). Slinky in hand: Physics & waves science activity. *Science Snacks*. Accessed Jun 3, 2024, https://www.exploratorium.edu/snacks/slinky-in-hand/print

Fančovičová, J., & Prokop, P. (2019). Examining secondary school students' misconceptions about the human body: Correlations between the methods of drawing and open-ended questions. *Journal of Baltic Science Education, 18*(4), 549–557.

Gamarra, E., & Nuygen, T. (2024). Spatial language as a tool to improve spatial ability. In R. M. Klapwijk, & G. Duffy (Eds.), *Making space for spatial thinking in STEAM education, teacher professional development modules*. TU Dublin and TU Delft. www.sellstem.eu

Gentner, D. (2016). Language as cognitive tool kit: How language supports relational thought. *American Psychologist, 71*(8), 650–657.

Gifford, S., Gripton, C., Williams, H., Dr, Lancaster, A., Bates, K. E., Williams, A. Y., Gilligan, K. A., Borthwick, A., & Farran, E. K. (2022). *Spatial reasoning in early childhood*.

Gilligan, K. A., Hodgkiss, A., Thomas, M. S., & Farran, E. K. (2019). The developmental relations between spatial cognition and mathematics in primary school students. *Developmental Science, 22*(4), Article e12786.

Hambrick, D. Z., Libarkin, J. C., Petcovic, H. L., Baker, K. M., Elkins, J., Callahan, C. N., Turner, S. P., Rench, T. A., & LaDue, N. D. (2012). A test of the circumvention-of-limits hypothesis in scientific problem solving: The case of geological bedrock mapping. *Journal of Experimental Psychology: General, 141*(3), 397–403.

Hawes, Z. C., Gilligan-Lee, K. A., & Mix, K. S. (2022). Effects of spatial training on mathematics performance: A meta-analysis. *Developmental Psychology, 58*(1), 112.

Hodgkiss, A., Gilligan, K. A., Tolmie, A. K., Thomas, M. S. C., & Farran, E. K. (2018). Spatial cognition and science achievement: The contribution of intrinsic and extrinsic spatial skills from 7 to 11 years. *British Journal of Educational Psychology, 88*, 675–697.

Hubber, P., Tytler, R., & Haslam, F. (2010). Teaching and learning about force with a representational focus: Pedagogy and teacher change. *Research in Science Education, 40*, 5–28.

Jansen, Y., Dragicevic, P., Isenberg, P., Alexander, J., Karnik, A., Kildal, J., Subramanian, S., & Hornbæk, K. (2015). Opportunities and challenges for data physicalization. In *proceedings of the 33rd annual acm conference on human factors in computing systems* (pp. 3227–3236).

Jornet, A., & Roth, W. M. (2018). Imagining design: Transitive and intransitive dimensions. *Design Studies, 56*, 28–53.

Julià, C., & Antolì, J. Ò. (2018). Enhancing spatial ability and mechanical reasoning through a STEM course. *International Journal of Technology and Design Education, 28*(4), 957–983.

Kell, H. J., Lubinski, D., Benbow, C. P., & Steiger, J. H. (2013). Creativity and technical innovation: Spatial ability's unique role. *Psychological Science, 24*(9), 1831–1836.

Klapwijk, R., & Stables, K. (2023). Design learning: Pedagogic strategies that enable learners to develop their design capability. In D. Gill, D. Irving-Bell, M. McLain, & D. Wooff (Eds.), *Bloomsbury handbook of technology education* (pp. 271–289). Bloomsbury Publishing.

Lauer, J. E., Yhang, E., & Lourenco, S. F. (2019). The development of gender differences in spatial reasoning: A meta-analytic review. *Psychological Bulletin, 145*(6), 537–565.

Liebschner, J. (2006). *A child's work: Freedom and guidance in Froebel's educational theory and practice.* Lutterworth Press.

Lowrie, T., Logan, T., Harris, D., & Hegarty, M. (2018). The impact of an intervention program on students' spatial reasoning: Student engagement through mathematics-enhanced learning activities. *Cognitive Research: Principles and Implications, 3*, 1–10.

Maresch, G., & Sorby, S. A. (2021). Perspectives on spatial thinking. *Journal for Geometry and Graphics, 25*(2), 271–293.

Mishra, R., & Klapwijk, R. M. (2024). 'Bear hunt' sparks change: Using lesson study to integrate spatial thinking in early childhood design education. In *German Conference on Spatial Cognition* (pp. 131–146). Springer Nature Switzerland.

National Research Council. (2006). *Learning to think spatially.* The National Academies Press.

National Research Council. (2012). *A framework for K-12 science education: Practices, crosscutting concepts, and core ideas.* National Academies Press.

Neuburger, S., Ruthsatz, V., Jansen, P., & Quaiser-Pohl, C. (2015). Can girls think spatially? Influence of implicit gender stereotype activation and rotational axis on fourth graders' mental-rotation performance. *Learning and Individual Differences, 37*, 169–175.

Newcombe, N. (2017). Harnessing spatial thinking to support stem learning (OECD Education Working Papers No. 161). OECD Publishing.

Newcombe, N. S., & Frick, A. (2010). Early education for spatial intelligence: Why, what, and how. *Mind, Brain, and Education, 4*(3), 102–111.

Newcombe, N. S., & Shipley, T. F. (2014). Thinking about spatial thinking: New typology, new assessments. *Studying visual and spatial reasoning for design creativity* (pp. 179–192). Springer, Netherlands.

Oxford Primary Illustrated Science Dictionary. (2019). Oxford University Press. ISBN: 978–0–19–277246–6.

Poelman, Y. (2015). *De natuur als uitvinder; Miljarden jaren aan innovatie gratis beschikbaar.* Carrera.

Ramey, K. E., & Uttal, D. H. (2017). Making sense of space: Distributed spatial sensemaking in a middle school summer engineering camp. *Journal of the Learning Sciences, 26*(2), 277–319. https://doi.org/10.1080/10508406.2016.1277226

Sonneveld, L., Klapwijk, R. M., & Stappers, P. J. (2024). Constructing and storytelling: Accommodating different play orientations in learning spatial thinking. In *Frontiers in Education* (Vol. 9, p. 1307951). Frontiers Media SA.

Sonneveld, L., Stappers, P. J., Heijne, K. G., Klapwijk, R. M. (2023). *Accomodating individual play preferences of preschoolers in design education focused on spatial ability* (Master thesis). Delft University of Technology.

Sorby, S. A. (2009). Educational research in developing 3-D spatial skills for engineering students. *International Journal of Science Education, 31*(3), 459–480.

Strand, I., & Lutnæs, E. (2022). Developing spatial skills through design of built environments: Art and crafts educators'strategies. *Design and Technology Education: An International Journal, 27*(3), 36–57.

Uttal, D. H., & Cohen, C. A. (2012). Spatial thinking and STEM education: When, why, and how? *Psychology of learning and motivation* (Vol. 57, pp. 147–181). Academic Press.

Uttal, D. H., Meadow, N. G., Tipton, E., Hand, L. L., Alden, A. R., Warren, C., & Newcombe, N. S. (2013b). The malleability of spatial skills: A meta-analysis of training studies. *Psychological Bulletin, 139*(2), 352.

Uttal, D. H., Miller, D. I., & Newcombe, N. S. (2013a). Exploring and enhancing spatial thinking links to achievement in science, technology, engineering, and mathematics? *Current Directions in Psychological Science, 22*(5), 367–373.

Van Mechelen, M. (2016). Designing technologies for and with children: Theoretical reflections and a practical inquiry towards a co-design toolkit.

Wai, J., Lubinski, D., & Benbow, C. P. (2009). Spatial ability for STEM domains: Aligning over 50 years of cumulative psychological knowledge solidifies its importance. *Journal of Educational Psychology, 101*(4), 817.

Zhu, C., & Klapwijk, R. (2024). The spatial aspect of designing: Opportunities, challenges, and conjectures on engaging pupils in spatial thinking through design education. In *German conference on spatial cognition* (pp. 97–113). Springer Nature Switzerland.

Zhu, C., Klapwijk, R., Silva-Ordaz, M., Spandaw, J., & de Vries, M. J. (2023b). Cognitive and embodied mapping of data: An examination of students's spatial thinking in data physicalization. *Frontiers in Education* (Vol. 8, p. 1308117).

Zhu, C., Klapwijk, R., Silva-Ordaz, M., Spandaw, J., & de Vries, M. J. (2024). Investigating the role of spatial thinking in children's design ideation through an open-ended design-by-analogy challenge. *International Journal of Technology and Design Education*, 1–30.

Zhu, C., Leung, C. O. Y., Lagoudaki, E., Velho, M., Segura-Caballero, N., Jolles, D., Duffy, G., Maresch, G., Pagkratidou, M. & Klapwijk, R. (2023a). Fostering spatial ability development in and for authentic STEM learning. *Frontiers in Education* (Vol. 8, p. 1138607).

Chapter 5
Developing Students' Science Understanding with Meaningful Language Activities and Semantic Gravity

Anna-Vera Meidell Sigsgaard

Abstract Learning science presents challenges for students, especially those learning in a second language. This is because second language learner students must grasp new scientific concepts while simultaneously mastering the language of instruction. Science lessons involve various linguistic demands, such as sharing observations, comparing data, and arguing with evidence. Teachers who fail to recognize the connection between knowledge and language may not provide adequate language support for additional language learners. A recent three-year research project in Copenhagen, Denmark, involved science and language teachers developing language activities for upper-primary years science lessons. In the project teachers explored the concept of *semantic gravity* from Legitimation Code Theory (LCT) and language activities informed by systemic functional linguistics (SFL). The project aimed to support teachers in (re)designing science units to enhance students' science literacy and their ability to communicate scientifically. Analysis of lesson observations using semantic gravity revealed that while students were engaged and excited by practical experiments, they often had difficulty translating these experiences into generalizable scientific knowledge. In response, teachers developed activities focusing on supporting students in creating less context-dependent meanings and using more scientific language in both oral and written forms. These activities aimed to improve students' ability to express scientific understanding, enhancing their communication of scientific ideas. This chapter includes examples of these activities, demonstrating how teachers can better support language development in science education.

A.-V. Meidell Sigsgaard (✉)
Department of Teacher Education, University College of Copenhagen, Copenhagen, Denmark
e-mail: anms@kp.dk

S. Earle et al. (eds.), *Primary Science Learning for Children, Teachers, and Communities*, https://doi.org/10.1007/978-981-95-0561-6_5

5.1 Vignette

Students in a fourth-grade science class are conducting experiments in pairs at the school's science lab. Each pair has a tub of water, aluminum foil, popsicle sticks, plastic straws, spoons, and metal nails. One pair is constructing an aluminum foil boat:

A: okay! then we're gonna

B: should we somehow have it, like

A: a little edge, right?

B: yeah. so it's not, like, so gigantic

A: before, there was hardly any room there

B: and then, yeah, then fold…

They quickly build their boat and start loading it with nails, counting aloud. Nearby, someone shouts, "Look, it floats!". The pair gets excited as they add more nails:

A: three! three-three-three!!

B: that's our record!

A: four!

B: five?

Both: Aaaaaw!! [as the boat sinks at six nails]

After cleanup, the teacher gathers the class to discuss their findings. Interested in consolidating students' understandings of how different constructions can affect a boat's ability to float she asks, "What worked … what are some good tips for building a boat?" Many students are eager to answer, reaching their hands up, wanting to tell her about each of their boats in turn and how many nails it could hold. Upon nomination, one student responds:

Student: yeah, well, it was ONLY able to hold, um, 96 nails.

[…]

Teacher: wow! that's wild!

why do you think it could carry so many nails?

Student: we used those popsicle sticks to make the walls.

then the walls were a little stronger,

and then we also had a lot of aluminum foil

Teacher: so the shape,

the thing with the walls

it actually makes a difference!

Student: and then there were also popsicle sticks underneath

In her responses to each student, the teacher prompts them to think about why they got the results they got. She also asks for more specifics about each construction as well as which adjustments students made in their following versions. Sometimes she adds observations, drawing their attention to what might affect their result. Despite students' enthusiasm about their experiments and the teacher's prompts, however, many students struggle to articulate what makes for a good boat design or explain the scientific principles behind their observations.

This scenario highlights a common challenge in science education: bridging the gap between hands-on activities and conceptual understanding. While students are engaged and excited by practical experiments, they often find it difficult to translate these experiences into generalizable scientific knowledge. This disconnect underscores the need for targeted language support in science education, especially for students learning in a second language.

This chapter explores how integrating language development with science content can enhance students' ability to communicate scientifically and deepen their understanding of complex concepts. By introducing the concept of *semantic gravity* from Legitimation Code Theory, this chapter offers science educators a practical framework for designing activities that scaffold students' progression from concrete experiences to abstract scientific principles. This approach not only supports students in developing scientific literacy but also equips them with the language skills necessary to articulate their understanding effectively, thereby making science education more accessible and meaningful for all learners.

5.2 Introduction

This chapter reports on findings from a research and development project concentrating on upper-primary science teaching in Danish public schools. The project aims to address commonly known challenges of science teaching: how to ensure that all students learn to transfer experiences from *doing* hands-on science activities, to *understanding* the scientific concepts those activities are meant to demonstrate (Musharrat, 2020). Participating in experiments and doing hands-on activities does not in itself ensure that students understand the scientific content in focus, nor necessarily that they learn to communicate about it actively (Leach & Scott, 2003; Lemke, 1998). In fact, many students in primary school science often have difficulties in accessing and using appropriate scientific language. This can be attributed to the many speaking and listening demands embedded in scientific discourse, such as when students are expected to share observations, compare data, critiquing findings, arguing with scientific evidence and developing new perspectives (Jacobsen & Mulvad, 2022; Polias, 2016). For second-language learners these challenges are multiplied, as they are most often both learning the new scientific concepts as well as the language of instruction at the same time (Gee, 2008; Gibbons, 2003; O'Hallaron et al., 2015).

When teachers do not see knowledge and language as fundamentally connected and interdependent, they may not provide students with the necessary language support in their lessons. Therefore, as part of effectively teaching their students to communicate and understand scientific concepts, teachers must also support them in developing specialized scientific vocabulary relevant to the topic as well as more general language skills relevant to communicating scientifically. The project reported on here focuses on students, their talk, how different activities require different kinds of talk, and how different kinds of talk connect to their understandings of science.

5.3 Challenges in Upper-Primary years Science from the Perspective of Language

The project reported on here, focuses on the teaching of science in the upper-primary grades in Denmark. At this stage of schooling, children are between the ages of 9 and 12. Generally speaking, students in upper-primary years of schooling experience new demands on their language use both receptive (when reading texts and listening to subject teachers) and productive (when they communicate in each subject, orally and in writing). In these grades, students go from learning how to read texts to reading texts for the purpose of learning new content (Gibbons, 2009; Martin & Rose, 2012). At the same time, the texts in these grades become more demanding and difficult to understand (Christie, 1998; O'Hallaron et al., 2015; Schleppegrell, 2004). As the topics become more abstract, students increasingly encounter meanings where the described scientific phenomenon does not necessarily have an equivalent in their experienced world (Lee et al., 2013; Mork, 2016).

Science teaching in the upper-primary grades becomes more challenging as the subject language becomes more abstract, more technically specific, and increasingly begins to reflect the scientific registers and linguistic characteristics of natural science (Christie & Derewianka, 2008; Jacobsen & Mulvad, 2022). This shift poses challenges for second language learners in particular (Leach & Scott, 2003; Rose, 2011). Research referring to "the fourth-grade slump" (Chall et al., 1990) shows amongst other things that second language learners in particular fall behind around this grade level (Møller et al., 2012) and can have difficulty performing at the academic level expected (Andersen et al., 2012; Beuchert et al., 2020; Collier & Thomas, 2007; Martin, 1990). Language and communication thus play an increasingly decisive role in creating understandings of scientific topics, and in students' ability to participate in classroom discussions (Gibbons, 2003; Polias, 2016; Windschitl et al., 2018).

Teaching science together with attention to the language of science is critical to the development of students' literacy both in terms of science literacy and language literacy (Carrejo & Reinhartz, 2014; Commins & Miramontes, 2006). Strategies for developing science literacy and language literacy together support both learning science and the language of science (Collier, 1998; Collier & Thomas, 2007; Commins & Miramontes, 2006). For this to be successful, teachers' awareness of the role of meaningful dialogue is key in supporting literacy development (Halliday, 1994; Painter, 1996). International research and development work points out that a functional understanding of language (Lee, 2005; Martin & Rothery, 1980) along with a conceptualization of the scientific content (Blackie, 2014; Georgiou et al., 2014; Macnaught et al., 2013; Mouton & Archer, 2019) can offer effective pedagogic approaches, and give more students access to the academic material.

5.4 Project Design and Project Background

Inspired by design-based research (Barab & Squire, 2004), the project described here incorporated existing research-based findings on language pedagogy and science teaching in a Danish upper primary school science context. The aim was to introduce the participating teachers to analytical and pedagogic tools useful for identifying students' challenges with understanding and communicating effectively about scientific understandings. These tools were then used to redesign lessons in selected units of study through a series of interventions. To do this, it was important to create space for collaborative reflection and planning within the context of the teachers' lessons.

During the project's first year (school year 2021–22), teacher-educators/researchers worked closely with one pair of teachers in the project's first school to develop and incorporate meaningful language activities for their upper-primary science lessons. In the following school year two more teacher-pairs at three new schools joined the project to further develop the activities in the context of their own lessons. The teachers' participation was enabled by covering costs of substitute teachers during the lessons the participating teachers would miss during project participation including joint observations, reflection/planning sessions, seminars as well as participation in the yearly Big Bang science education conference held in Denmark.

The project interventions had their starting point in the teachers' annual cycle and the topics and materials the teachers already had planned to use. Three units of study (science topics) for each school year were chosen by the science teacher for intervention. Each of these topics was then jointly planned with teachers and researchers in sessions in and outside the school. The planning and reflection sessions facilitated joint reflection making space for the teachers' rich understandings of practice in their contexts. Theoretical knowledge and pedagogic understanding was brought to the sessions by the teacher-educators. The combination of teachers' practical knowledge and researchers' theoretical knowledge for reflecting on specific video excerpts created a collaborative and trusting space between researchers and teachers. The reflection and planning sessions included joint planning, reviewing lessons and adjusting future lesson plans based on reflections made during the sessions. Three times a year, all the projects' teachers and researchers/teacher-educators met for a three-day seminar in which they shared their experiences incorporating linguistic work for supporting students' scientific understandings and planned their next topics incorporating these reflections.

For each topic planned, data were collected in the form of audio recordings of planning/reflection sessions and observations from lessons taught. Most lessons were observed and video recorded both on a stationary camera and on hand-held devices.

This allowed for recording the way the lesson unfolded from a whole-class perspective, as well as to focus on individual and group work during the lessons. To facilitate reflections on the language students were using in the observed lessons, the researchers chose excerpts from the video observations for collective reflection in the following sessions. Given the project's focus, these excerpts often showed the students' and teachers' language-use in different activities, but also included examples of students working in different situations and settings. The reflection sessions often also included adjusting activities in the upcoming lessons to increase the students' engagement with the language of the science being learned. Supplemented with statements from interviews and the reflection meetings, examples of student–student and student–teacher interactions from the projects' first school, constitute data for this chapter. Thus, the overarching research question for the project was:

> How can teachers support students in developing their science understandings with attention to the language of the subject?

5.5 Developing Attention to Scientific Language and Scientific Knowledge with the First Team of Teachers

The project's first intervention began with the first school's team of two experienced teachers: Pernille, a passionate science teacher, and Kamille, skilled in inclusive teaching and language arts. Both teachers have many years' teaching-experience in primary school and have been teaching together as a team for several years. At the start of the project (in 2021) they were teaching their school's two fourth grade classes, covering the majority of the classes' subjects between the two of them.

Their initial unit of study on 'seafaring and sailing' lasted six weeks, with a block of two back-to-back 90-min lessons per week. Activities included an excursion to the nearby city harbor, where students took pictures of different kinds of nautical vehicles. In a following lesson, students' photos were printed out and students were asked to work together to categorize their photos. During the unit, students also participated in hands-on activities meant to explore sinking/floating phenomena, as described in the vignette at the beginning of this chapter. Throughout the unit the teachers also led whole-class discussions meant to consolidate students' experiences and understandings.

A key reflection session occurred when the teachers viewed video clips from their lessons. These had been chosen by the researchers, as they showed two distinct language-use situations: 1. Students working on hands-on experiments (building aluminum foil boats), and 2. a class discussion led by Pernille. Showing the teachers the video clips of the two language-use situations (also described in the vignette) was a useful starting point for them to see how the language demands on the students can be very different as dictated by the activity.

While watching the video of the students building an aluminum boat, the teachers noticed that although the students were highly engaged, they weren't using any scientific vocabulary. Unsurprisingly, the language the students use during their hands-on activities is spoken-like, dialogic, and accompanying the students' immediate actions. Students are talking about '*it*', '*a little edge*', and '*there*'. The context of the situation means the students do not need specific words or vocabulary (such as *sink, float, surface area, weight, shape, strength, buoyancy*) to build their boat or to get their meaning across. Being asked to reflect on the language students used in this video, the teachers recognized that this informal language was natural for the hands-on context and how the activity itself did not require specific use of scientific or topic-specific vocabulary.

At the same time watching this video supported the teachers' and researchers' commonly held view that hands-on activities are important to the development of students' understanding of the scientific content. The experience of building the boat and observing what happens to it when they load it with nails and when it sinks gives each collaborating group of students a shared experience. Any learning about buoyancy they are making here, however, is closely related to their specific experience. For this reason, asking students to gather and share their findings is an important aspect of helping them consolidate their learning.

Upon seeing the second video-clip of the class discussion, the teachers observed that the students struggled to articulate general principles about buoyancy, instead focusing on specific details of their constructions. They noticed that student engagement in this activity is different from while they were building their boats: only one student speaks at a time while the rest of the class must listen. From the point of view of the language being used, the student speaking uses more specific words to describe their own constructions (ie. '*We used those popsicle sticks to make the walls* […] *and then we also had a lot of aluminum foil*'). However, in terms of articulating general findings or observations about what makes something float, most students appear to need support. Only with prompts from the teacher do they try to formulate reasons for why things had turned out the way they had (ie. '*then the walls were a little stronger*').

These observations provided an entry point to introduce teachers to the concept of *semantic gravity* as a way of understanding the different levels of context-dependence of knowledge in their lessons. Stemming from Legitimation Code Theory, the concept of semantic gravity is used to make visible the context-dependence of knowledge (Maton, 2014). Semantic gravity refers to the degree to which meaning is context-dependent and can be stronger or weaker along a continuum of strengths (Maton, 2014). When semantic gravity is stronger (SG +), meaning is closely tied to its context, and conversely, when semantic gravity is weaker (SG–), meaning is less dependent on context (ibid.). Inspired by Kirk (2017), Meidell Sigsgaard and Jacobsen (2021) and Macnaught (2021) in the context of academic writing, and Georgiou (2016) in the context of tertiary Science education, we operate with three strengths of semantic gravity within the frame of this project, see Fig. 5.1, below.

Reflecting the way that more *abstract* meanings are typically placed higher than more *concrete* meanings in many visualizations, stronger semantic gravity (SG+) is at

	Coding categories for strengths of semantic gravity	Description of coded content	Examples from the data
semantic gravity	Weaker semantic gravity (SG-)	Meanings expressing a scientific concept or understanding without referencing a specific situation.	"A thing can float if it can displace more water than it weighs in itself"
	Medium semantic gravity (SG0)	Meanings expressing general tendencies and/or explanations	"It was better [floated, *ed.*] with the bigger surface"
	Stronger semantic gravity (SG+)	Descriptions of the conducted experiments/specific contexts in the classroom	"our boat sank when we put too many screws on it"

Fig. 5.1 Three levels of semantic gravity in the context of upper-elementary science lessons

the bottom and weaker semantic gravity (SG-) is at the top of the figure. To make the concept more accessible to the teachers, we followed Meidell Sigsgaard and Jacobsen (2021) in color-coding the three levels with green for strongest SG+ (conceptualizing green grass), yellow for the middle strength of semantic gravity SG0 (conceptualizing yellow flowers blooming), and blue for weakest SG- (conceptualizing the blue sky).

Using this framework, the teachers and researchers together analyzed the approximately six-minutes of classroom discussion from the whole-class discussion in video-clip 2. In the six minutes of conversation, 40 'turns' were taken by teachers (T) and one student at a time (S). Each turn was numbered (and divided into clauses), showing who was speaking, see Fig. 5.2.

The analysis revealed that most student responses remained at the strongest level of semantic gravity (SG+), closely tied to each of their specific experiments. In the video-clip, only the teacher makes one meaning at the weakest semantic gravity level (SG-) in turn 39. In this turn she responds to a student's attempted explanation in turn 38 in which he says '*we used those popsicle sticks to make the walls* [SG+] *then the walls were a little stronger* [SG0], *and we had a lot of aluminum foil* [SG+]'. Pernille rewords this explanation saying, '*so the shape*' (in turn 39), as a way of drawing the students' attention to an understanding of how an object's shape can impact its ability to float – a characteristic that can affect all objects' buoyancy. In doing so, she makes a less context-dependent meaning, at the weakest strength of semantic gravity.

Analyzing the video confirmed the teachers' first impression from watching the video that students struggled to explain *why* a particular construction was better than another. The teachers also realized that although explanations and hypotheses about buoyancy were being made throughout the conversation, these were almost exclusively contributed by the teacher, not the students. In other words, based on the classroom discussion alone, teachers could not deduce whether students had actually understood that the shape of an object would affect its buoyancy. This analysis led to several key insights for the teachers.

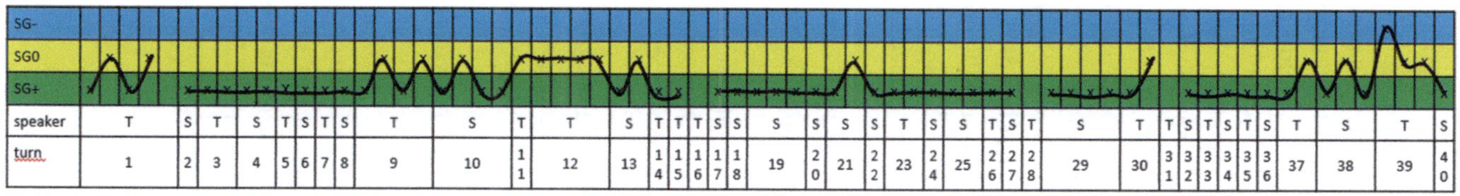

Fig. 5.2 Semantic gravity in 40 turns during a classroom discussion with the teacher (T) and students, one at a time (S) (Adapted from Schjøtt et al., 2024)

By watching the two video excerpts, they recognized how activity design significantly influences language use. While the hands-on activities were engaging and promoted active participation, they didn't naturally elicit the use of scientific language. Students were deeply involved in the practical aspects of building and testing their boats, but this didn't automatically translate into being able to discuss the scientific ideas which the activities were meant to illustrate.

Pernille realized the need for more scaffolding to help students move from describing their specific experiences to articulating general principles about buoyancy. She noticed that even when prompted by her, students struggled to explain why certain boat designs worked better than others, often ending up focusing on the number of nails their boat could bear rather than the underlying scientific concepts. Kamille, drawing on her language teaching expertise, suggested that more attention to scientific vocabulary before the discussion might have helped students express their ideas more scientifically. She observed that while students were using more specific words to describe their constructions in the class discussion (e.g. '*popsicle sticks*' and '*aluminum foil*') as compared to when building the boats, they weren't incorporating the scientific terms related to buoyancy and flotation they had learned earlier. In other words, students were not connecting the practical work to the scientific knowledge in focus.

These observations highlighted the need for including strategies to ensure broader participation in discussions to help all students develop their ability to articulate scientific ideas. Viewing their lessons from the perspective of semantic gravity, both teachers recognized the importance of intentionally incorporating activities that support students in making less context-dependent meanings. They saw the need to bridge the gap between the hands-on experiences and the more abstract scientific concepts they wanted students to grasp.

Insights from the collaborative reflective sessions provided a foundation for the teachers to rethink their approach to language use in science lessons and motivated them to explore new pedagogical strategies in subsequent units. To support them in introducing students to relevant language features in future science lessons, the researchers introduced aspects of genre-pedagogic approaches (Brisk, 2015; Derewianka & Jones, 2016; Gibbons, 2015; Halliday & Matthiessen, 2014; Polias, 2016). Reading texts such as Derewianka (1990) and excerpts from Polias (2016) and Jacobsen and Mulvad (2022) together with the researchers providing the teachers with specific ideas for incorporating activities which integrate reading and writing while engaging meaningfully with the content knowledge being taught. These activities were specifically chosen with the intention of supporting students in learning to make meanings in the two higher levels of semantic gravity, as demonstrated in the following section of this chapter.

5.6 Targeted Language Activities to Scaffold Students in Producing Less Context Dependent Meanings

The second year of the project saw the implementation of targeted language-based activities across three different schools, each designed to address specific aspects of scientific language development and conceptual understanding. These activities were crafted to encourage students to operate at weaker levels of semantic gravity (SG0), providing an approach to scaffolding students' scientific literacy. The following section highlights three key activities that exemplify the practical application of semantic gravity in science education. The first activity focuses on raising students' awareness of scientific and general academic vocabulary, aiming to scaffold their ability to 'speak like a scientist.' The second, termed 'puzzle sentences,' engages students meaningfully with scientific texts, helping them to bridge the gap between everyday language and scientific discourse. The final activity demonstrates the importance of slowing down conversations, providing second language learners with crucial thinking time to experiment with different ways of expressing scientific concepts. Implementing these activities in science lessons not only allows students to build meta-language skills but also enhances their awareness of how language functions to represent more abstract and complex scientific meanings. Together, these activities showcase the potential of integrating language development with scientific content.

5.6.1 Activity 1: Raising Awareness of Scientific, Everyday, and General Academic Vocabulary

At the beginning of a unit on the water cycle, a fourth-grade science teacher, Nina, implemented an activity designed to develop students' metalinguistic awareness by distinguishing between different types of vocabulary. The activity began with a class brainstorming session, where students shared everything they already knew about water and the water cycle. Drawing from this initial discussion, the teacher carefully selected a list of key words essential for understanding and communicating about the water cycle.

In the following lesson, the teacher presented these words to the class and asked students to categorize them as 'everyday,' 'scientific,' or 'in-between' vocabulary. This task proved to be a powerful tool for engaging students with both the scientific content and the language used to discuss it. As students worked on categorizing the words, rich discussions emerged about why certain words belonged in each category. They discovered that some words, like 'analyze,' 'compare' and 'hypothesis,' didn't fit neatly into either the everyday or scientific categories, justifying the 'in-between' category. These conversations revealed that students' categorizations were influenced by their personal experiences and familiarity with the words, as well as the context in which they were used. Paying attention to the vocabulary types across several units of study, students noticed that the in-between words could often be reused across different science units and even in other subjects.

Reflecting on the different vocabulary types not only helped students identify scientific terminology but also made them aware of how everyday words can take on specific meanings in scientific contexts, and how general academic vocabulary is crucial for understanding and participating in scientific discourse. For example, students realized that while 'argument' is an everyday word, in science it takes on a more specific meaning related to evidence-based reasoning. Meanwhile the teachers recognized these as general academic words needed across subjects, and therefore the importance of highlighting these words and ensuring students' understanding of these.

By explicitly focusing on these different types of vocabulary, the teacher created opportunities for students to 'speak like a scientist,' improving their scientific literacy and their ability to engage with the topic of the water cycle more effectively. This approach was particularly beneficial for the many second language learners in the class, as it provided a structured way to build both their scientific understanding and their language skills simultaneously, right at the outset of the unit. Moreover, it highlighted the importance of general academic vocabulary, which is often overlooked in science instruction but is crucial for overall academic success.

5.6.2 Activity 2: Puzzle Sentences

The second activity, 'puzzle sentences,' was implemented during a fourth-grade science unit on hearing and sound. This unit explored how humans produce and perceive sound, including the mechanics of the ear and the process of sound production in the human body. The activity aimed to make a challenging scientific text about sound production more accessible to students.

For this activity, Pernille identified a text from the students' textbook which she wanted students to read. Kamille, however, was sure this text would be too difficult for the students to understand without scaffolding. The original text read:

> When we talk, sing and shout, we make the air vibrate. We call these sound waves. We make sounds with air from the lungs, the two vocal cords in the larynx and the entire oral cavity. When we talk, shout or sing, we tighten the vocal cords. The airstream from the lungs causes the vocal cords to vibrate. Then the air vibrates in the entire oral cavity. Sounds are formed. (Dalgaard & Rahbek, 2005: 6B, *own translation*)

The project researchers suggested breaking down the text into functional units using transitivity analysis from systemic functional linguistics (Halliday & Matthiessen, 2014). Resulting puzzle pieces could be 'things/phenomena' in the text (e.g. '*we*', '*soundwaves*', '*the airstream from the lungs*' and '*vocal cords*'), the processes that these 'things' participate in (e.g. '*vibrates*', '*call*', '*tighten*' and '*are formed*'), the circumstances under which these processes happen (e.g. '*when we talk, sing and shout*' and '*in the entire oral cavity*') as well as logical connectors (e.g.

'*then*'). A possible sentence out of some of the puzzle pieces described here could be: *the vocal cords causes the airstream from the lungs to vibrate.* However, this doesn't make sense in terms of the science, as it is not the vocal cords that cause the airstream to vibrate, but rather the airstream causes the vocal cords to vibrate. Arranging the puzzle pieces into meaningful sentences, therefore, requires engaging with and understanding the content.

Students worked in pairs to reconstruct meaningful sentences from these puzzle pieces, talking about what they were doing as they went:

> Boy: ...in capital letters. Yeah, here's a capital one
>
> Girl: Okay.. Maybe something like we make sounds ... could this fit?
>
> Boy: This one ... when we talk, shout, or sing
>
> Girl: Or it could also be we make sounds throughout the oral... the oral cavity
>
> Boy: The airflow from the lungs .. causes the vocal cords to vibrate. Easy-peasy lemon squeezy! (*my translation from Danish to English*)

An important observation from this interaction is that the boy and the girl use different strategies to construct sentences. The girl has a good understanding of the content, and engages directly with each puzzle piece's meaning, in trying to put pieces together that make sense, constructing the sentence, '*we make sounds throughout the entire oral cavity*'. The boy in this example, however, is less secure in his understanding of the phenomenon, but is still able to construct sentences that make sense and are most likely more complex than what he would create on his own. He uses strategies based on his knowledge about sentences, such as identifying pieces with capital letters and pieces with periods, and places these at the beginning or end of potential sentences. By doing this he puts suggests the sentence, '*The airflow from the lungs causes the vocal cords to vibrate.*' His partner, the girl, confirms that this makes sense with a look and a nod, while continuing with her own sentence-construction. His resulting sentence uses scientific terminology and expresses a cause-effect relationship. It is also a general statement of an observable (scientific) phenomenon rather than e.g. a direct reference to his own experiences. Having put together puzzle pieces and read aloud the resulting sentence, the boy ends up constructing an accurate sentence which he probably would not have made otherwise. This demonstrates a key strength of the activity: even students relying on strategies for sentence construction end up producing meanings about the scientific content that are generalizable.

Having created sentences with the puzzle pieces, the students were then asked to write their sentences down in their logbooks. Two different groups produced the following results:

1. The airstream from the lungs causes the vocal cords to vibrate. When we speak, shout or sing, we tighten the vocal cords. Sounds are formed in the entire oral cavity. We make sounds with air from the lungs, the two vocal cords in the larynx and the entire oral cavity.
2. The airstream from the lungs causes the vocal cords to vibrate. Then the air vibrates with air from the lungs, the two vocal cords in the larynx and the entire

> oral cavity. When we speak, shout or sing, we tighten the vocal cords. Sounds are formed. We make sounds in the entire oral cavity. (*My translation*)

Both of these students' logbook texts, while different from each other, make sense and demonstrate students' accurate understandings of the scientific content in focus.

The design of this activity, both the collaboration needed for putting together the puzzle sentence pieces, and then writing these sentences together to form a text in their logbook, required students to engage with scientific language directly while applying their understanding of how sounds are produced. In terms of semantic gravity, students are supported in moving from the stronger semantic gravity of their immediate experiences (like saying things such as '*I make sounds with my mouth*') to weaker levels of semantic gravity where they express more abstract scientific concepts, not directly related to themselves and their own experiences (i.e. '*The airflow from the lungs causes the vocal cords to vibrate*'). In this way, the design of the activity leads students to engage with scientific and academic language in a meaningful way. By constructing and then writing these sentences in their logbooks, students were able to produce more generalizable meanings about sound production, moving away from the stronger semantic gravity of their immediate experiences towards the weaker semantic gravity expressed by more abstract scientific concepts.

5.6.3 Insurance or Evaporation—Giving Students Time to Think

The third approach focuses on providing students with ample time to process information and experiment with different ways of expressing scientific concepts. This strategy was demonstrated particularly well during a focus group interview at the conclusion of a water cycle unit in a third-grade class. After completing the unit, students worked in small groups to demonstrate their understanding from the unit. The researcher presented a diagram of the water cycle, which students had previously labeled using key terms they could recall. The following interaction showcases how allowing students time to think and discuss can lead to deeper engagement with scientific vocabulary and concepts. When prompted to remember any additional terms, students collaboratively brainstormed, suggesting words such as 'groundwater,' 'seepage,' 'precipitation' and 'condensation.'

A particularly insightful moment occurred when a second language learner, struggling to remember a term, suggested 'forsikring' (insurance). While on the surface this suggestion seems irrelevant to the subject being discussed, it demonstrates the student's developing awareness of academic language patterns. By using the Danish '-ing' ending (which in English often corresponds to the '-tion' or '-age' endings of abstract nouns), the student showed she was noticing and trying to apply academic word formation rules, even if the base word was incorrect.

This moment led to a collaborative effort to understand and correct the term. Another student explained that 'forsikring' related to car insurance, while the original student clarified she was thinking of something related to 'nedsivning' (seepage), thinking about how water soaks through the earth down to the groundwater. This exchange highlights how giving students space to make mistakes and work through misunderstandings can foster language development alongside content understanding. The approach of slowing down conversations and allowing linguistic experimentation created an environment where students could navigate between everyday language and scientific terminology, practicing the use of the latter. By providing opportunities to discuss, question, and even make errors, this method supports second language learners in developing both their scientific knowledge and their ability to express it accurately in the target language.

5.7 Scaffolding Students' Scientific Understandings and Language Development

By thinking about what knowledge students are meant to engage with, the teachers in the project started designing units of study with the aim of expanding students' 'semantic range'. These units started with activities where the knowledge is hands-on and closely related to the context and progressing upwards (see Fig. 5.3) to weaker levels of semantic gravity. This section describes the final unit of study observed at the end of the project's second year of interventions in Pernille and Kamille's now 5th grade classes, which focused on the concept of insulation.

This unit showcases how the teachers' understanding of semantic gravity helped them plan a sequence of activities that systematically scaffolded students' development of academic language and scientific understanding. The unit began with a fair-test experiment (Activity 1) where students, working in groups, tested the insulating properties of different materials on bottles of hot water. This activity, situated at the stronger end of the semantic gravity continuum (SG+), allowed students to engage with the concept of insulation through direct experience. Following the experiment, the teachers facilitated a retelling activity (Activity 2), asking students to describe their experiments using everyday language and focusing on what materials they had chosen and what they did and when. In this activity, students rely on their experiences, setting the stage for the introduction of relevant and more scientific vocabulary.

As the unit progressed, the teachers included the first key activity described earlier: raising awareness of scientific, everyday, and general academic vocabulary (Activity 3). Key words students had used to describe their experiments were added to a word-cloud on the classroom whiteboard. In a whole-class discussion students were asked to place these words on a continuum from everyday to scientific language. This activity helped students develop a meta-language for talking about the language of science, a crucial step in their academic language development. Students then retold their experiments (Activity 4). This time, however, they were asked to describe their findings and remark on tendencies visible in their data and making sure to include words from the word cloud. In this way this retelling moved students towards a more

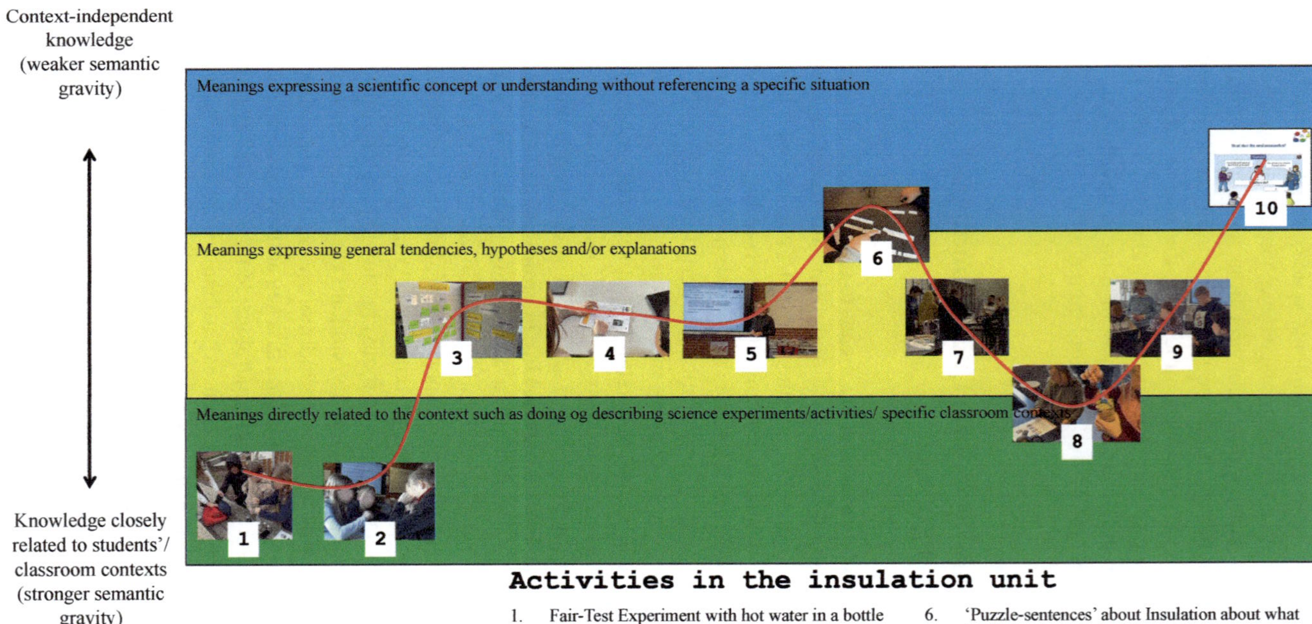

Activities in the insulation unit

1. Fair-Test Experiment with hot water in a bottle and two insulating materials
2. Retelling of the bottle experiment focusing on what students did
3. Reflection on everyday words and technical terms
4. Retelling of the experiment using data sheets and the word cloud
5. Joint-Construction of a Logbook Entry about the Water Bottle Experiment
6. 'Puzzle-sentences' about Insulation about what insulation is and what it's used for
7. Students Design a New Experiment with 3 Materials in Milk Cartons
8. Conduct the New Experiment with Milk Cartons and Collect Data
9. Input Data into Excel and Create Graphical Representation of Measurements from the Milk Carton Experiment
10. Concept discussion in new groups

Fig. 5.3 Activities from the last unit of study on insulation placed on the semantic gravity scale

scientific register while also drawing their attention to more abstracted thinking (tendencies in their data).

The second key activity, 'puzzle sentences' (Activity 6), was incorporated midway through the unit. Students worked in pairs to reconstruct sentences about insulation using provided text fragments. This activity, operating at a medium and weakest level of semantic gravity, encouraged students to engage with scientific texts while applying their growing understanding of insulation principles. Placing this activity after students have an operational understanding of the content makes it possible for them to engage with the texts at this point. By grappling with the language of science in a supportive context, students began to bridge the gap between their everyday language and scientific discourse.

As the unit neared its conclusion, students designed and conducted a new experiment testing the insulating properties of materials in milk cartons (Activities 7 and 8). This activity, building on their earlier experiences, pushed students to apply their evolving understanding of insulation to a novel context. Students then inputted their data into Excel and created graphical representations (Activity 9), further engaging with the scientific process and the language associated with data analysis. The teachers facilitated the third key activity, concept discussion (Activity 10), in small groups. Using a word cloud and sentence starters, students formed hypotheses about the insulating properties of different materials, employing the scientific language they had acquired throughout the unit. This activity, situated at the weaker end of the semantic gravity continuum (SG-), challenged students to articulate their understanding of insulation principles in more generalized terms, demonstrating their progress in both conceptual understanding and academic language proficiency.

By structuring the unit to move progressively 'up' from stronger to weaker semantic gravity, the teachers effectively scaffolded their students' development of scientific understanding and academic language, increasing students' semantic gravity range. The strategic incorporation of the three key activities at different points in the learning sequence, along with other scaffolding activities, supported students in bridging the gap between their everyday experiences and the abstract principles of insulation. This approach, grounded in an understanding of semantic gravity, offers a promising model for science educators seeking to integrate language development with content learning in their classrooms.

5.8 Conclusion

The integration of semantic gravity as an analytical tool in upper-primary science education offers a promising approach to addressing the challenges of developing both scientific understanding and academic language proficiency. As demonstrated through the experiences of Pernille and Kamille, viewing classroom activities through the lens of semantic gravity provides teachers with valuable insights into the relationship between hands-on experiences, language use, and students' conceptual understandings of the science content.

The challenge of bridging the gap between engaging hands-on activities and the development of generalizable scientific knowledge is a well-documented issue in science education (Hodson, 1990; Leach & Scott, 2003). This study highlights how conscious attention to the context-dependence of knowledge, as conceptualized through semantic gravity, can guide teachers in designing more effective learning sequences.

By structuring units of study to help students expand their semantic gravity range, moving from stronger to weaker semantic gravity, teachers can scaffold students' progression from concrete, context-dependent understandings to more abstract, generalizable scientific concepts. This approach addresses the often observed difficulty students face in translating their practical experiences into broader scientific principles. The use of semantic gravity as a planning tool also provides a framework for integrating language development with content learning, a crucial consideration given the increasing linguistic demands of upper-primary science education (Christie & Derewianka, 2008; O'Hallaron et al., 2015). This integrated approach is particularly beneficial for second language learners, who face the dual challenge of mastering both scientific concepts and the language of instruction.

The project reported on here demonstrates the value of collaborative reflection between teachers and researchers. The process of analyzing classroom interactions through the lens of semantic gravity not only enhanced the teachers' understanding of their students' learning processes but also motivated them to explore new pedagogical strategies. The observations and insights gained have been implemented in the researchers' teacher-education programs, helping to prepare future science teachers, and have the potential of being implemented in other teacher education programs as well.

Importantly, the impact of this approach on students was evident in their own reflections towards the end of the project. When interviewed about their science lessons after completing the last intervention, students provided insightful feedback. One student commented that initially, it was challenging to engage in so much writing. Another student agreed but added, 'But now we sound smarter.' This student perception underscores the importance of developing not only scientific understanding but also academic language proficiency, enhancing students' ability to articulate complex ideas.

While this study focused on upper-primary science education in Denmark, the principles of using semantic gravity to scaffold language and content learning have potential applications across various educational contexts and subject areas. Future research could explore the long-term impacts of this approach on students' scientific literacy and academic language proficiency, as well as its effectiveness in different cultural and linguistic contexts. The integration of semantic gravity as a guiding principle in science education offers a promising path for enhancing both scientific understanding and language development. By providing teachers with a tool to consciously navigate between concrete experiences and abstract concepts, this approach has the potential to make science education more accessible and meaningful for all students, regardless of their linguistic background. The students' own recognition of their improved articulation of scientific concepts serves as a powerful testament to the effectiveness of this approach.

References

Andersen, D., Jakobsen, V., Jensen, V. M., Nielsen, S. S., Zacho Pedersen, K. C., Petersen, D. S., & Thorsen, K. M. (2012). *Indsatser for tosprogede elever: Kortlægning og analyse*. SFI-Det Nationale Forskningscenter for Velfærd.

Barab, S., & Squire, K. (2004). Design-based research: Putting a stake in the ground. *The Journal of the Learning Sciences, 13*(1), 1–14. https://doi.org/10.1207/s15327809jls1301_1

Beuchert, L., Christensen, V. T., & Jensen, S. V. (2020). *PISA Etnisk 2018: PISA 2018 med fokus på elever med indvandrerbaggrund* (p. 134). VIVE—Det Nationale Forsknings-og Analysecenter for Velfærd. www.vive.dk

Blackie, M. A. L. (2014). Creating semantic waves: Using legitimation code theory as a tool to aid the teaching of chemistry. *Chemistry Education Research and Practice, 15*(4), 462–469. https://doi.org/10.1039/C4RP00147H

Brisk, M. E. (2015). *Engaging students in academic literacies genre-based pedagogy for K-5 classrooms*. Routledge. https://www.routledge.com/Engaging-Students-in-Academic-Literacies-Genre-based-Pedagogy-for-K-5-Classrooms/Brisk/p/book/9780415737043

Carrejo, D. J., & Reinhartz, J. (2014). Teachers fostering the co-development of science literacy and language literacy with English language learners. *Teacher Development, 18*(3), 334–348. https://doi.org/10.1080/13664530.2014.914564

Chall, J. S., Jacobs, V. A. E., & Baldwin, L. (1990). *The reading crisis: Why poor children fall behind*. Harvard University Press.

Christie, F. (1998). Learning the literacies of primary and secondary schooling. In F. Christie & R. Misson (Eds.), *Literacy and schooling* (pp. 47–73). Routledge. https://doi.org/10.4324/9780203019559

Christie, F., & Derewianka, B. (2008). *School discourse—Learning to write across the years of schooling*. Continuum.

Collier, V. (1998). *Promoting academic success for ESL students: Understanding second language acquisition for school*. Bastos Book Co.

Collier, V., & Thomas, W. (2007). Predicting second language academic success in English using the prism model. In J. Commins & C. Davison (Eds.), *International handbook of English language teaching* (pp. 333–348). Springer. https://doi.org/10.1007/978-0-387-46301-8_24

Commins, N., & Miramontes, O. (2006). Addressing linguistic diversity from the outset. *Journal of Teacher Education, 57*, 240–246. https://doi.org/10.1177/0022487105285591

Dalgaard, I., & Rahbek, J. L. (2005). *Kommunikation 2: Fra natur til teknik: 3.-4. klasse*. Malling Beck.

Derewianka, B. (1990). *Derewianka 1990 rocks in the head.pdf*. In R. Carter (Ed.), *Knowledge about language and the curriculum: The LINC reader* (pp. 197–215). Hodder and Stoughton.

Derewianka, B., & Jones, P. (2016). *Teaching languages in context*. Oxford University Press.

Gee, J. P. (2008). *Getting over the slump: Innovation strategies to promote children's learning*. The Joan Ganz Cooney Center at Sesame Workshop.

Georgiou, H. (2016). Putting physics knowledge in the hot seat: The semantics of student understandings of thermodynamics. In K. Maton, S. Hood, & S. Shay (Eds.), *Knowledge-building: Educational studies in legitimation code theory* (pp. 176–192). Routledge. https://doi.org/10.4324/9781315672342-19

Georgiou, H., Maton, K., & Sharma, M. (2014). Recovering knowledge for science education research: Exploring the "Icarus effect" in student work. *Canadian Journal of Science, Mathematics and Technology Education, 14*(3), 252–268. https://doi.org/10.1080/14926156.2014.935526

Gibbons, P. (2003). Mediating language learning: Teacher interactions with ESL students in a content-based classroom. *TESOL Quarterly, 37*(2), 247. https://doi.org/10.2307/3588504

Gibbons, P. (2009). *English learners, academic literacy, and thinking: Learning in the challenge zone*. Heinemann.

Gibbons, P. (2015). *Scaffolding language scaffolding learning: Teaching English language learners in the mainstream classroom* (Bd. 2). Heinemann Publishing. https://doi.org/10.1016/j.sbspro.2016.10.032%0A https://doi.org/10.1080/00131881.2019.1600376%0A https://www.sciencepublishinggroup.com/j/her

Halliday, M. A. K. (1994). The place of dialogue in children's construction of meaning. In R. B. Ruddell, M. R. Ruddell, & H. Singer (Eds.), *Theoretical models and processes of reading* (4. udg., pp. 70–82). International Reading Association.

Halliday, M. A. K., & Matthiessen, C. (2014). *Halliday's introduction to functional grammar*. Routledge.

Hodson, D. (1990). A critical look at practical work in school science. The *School Science Review, 71*(256), 33–40.

Jacobsen, G. K., & Mulvad, R. (2022). *Ind i naturfag—Fra tekster til naturfaglig kompetence*. Akademisk Forlag.

Kirk, S. (2017). Waves of reflection: Seeing knowledges in academic writing. In J. Kemp (Ed.), *EAP in a rapidly changing landscape: Issues, challenges and solutions* (pp. 109–118). Garnet Publishing.

Leach, J., & Scott, P. (2003). Individual and sociocultural views of learning in science education. *Science & Education, 12*(1), 91–113. https://doi.org/10.1023/A:1022665519862

Lee, O. (2005). Science education with English language learners: Synthesis and research agenda. *Review of Educational Research, 75*(4), 491–530. https://doi.org/10.3102/00346543075004491

Lee, O., Quinn, H., & Valdés, G. (2013). Science and language for English language learners in relation to next generation science standards and with implications for common core state standards for English language arts and mathematics. *Educational Researcher, 42*(4), 223–233. https://doi.org/10.3102/0013189X13480524

Lemke, J. L. (1998). Multiplying meaning: Visual and verbal semiotics in scientific text. In J. R. Martin & R. Veel (Eds.), *Reading science* (pp. 87–113). Routledge. https://www.researchgate.net/publication/246905867

Macnaught, L. (2021). Demystifying reflective writing in teacher education with semantic gravity. In C. Winberg, S. McKenna, & K. Wilmot (Eds.), *Building knowledge in higher education enhancing teaching and learning with legitimation code theory* (pp. 19–36). Routledge. https://doi.org/10.1080/13562517.2016.1160045

Macnaught, L., Maton, K., Martin, J. R., & Matruglio, E. (2013). Jointly constructing semantic waves: Implications for teacher training. *Linguistics and Education, 24*, 50–63. https://doi.org/10.1016/j.linged.2012.11.008

Martin, J. R. (1990). Literacy in science: Learning to handle text as technology. In F. Christie (Ed.), *Literacy for a changing world* (pp. 79–117). Australian Council for Autralian Research.

Martin, J. R., & Rose, D. (2012). *Learning to write, reading to learn: Genre, knowledge and pedagogy in the Sydney school*. Equinox.

Martin, J. R., & Rothery, J. (1980). *Write it right* (Working Papers in Linguistics). Linguistics Department, University of Sydney.

Maton, K. (2014) *Knowledge and knowers: Towards a realist sociology of education*. Routledge.

Meidell Sigsgaard, A. V., & Jacobsen, S. K. (2021). Semantiske bølger: Analytisk redskab og didaktisk princip. In S. K. Knudsen & L. Wulff (Red.), *Kom ind i sproget: Flersprogede elever i fagundervisningen* (2. udg., pp. 73–90). Akademisk Forlag.

Møller, J. S., Jørgensen, J. N., & Holmen, A. (2012). Polylingual development among Turkish speakers in a Danish primary school—A critical view on the fourth grade slump. *International Journal of Bilingual Education and Bilingualism, 17*(1), 32–54. https://doi.org/10.1080/13670050.2012.735221

Mork, S. M. (2016). Å diskutere som del af utforskende arbejd i naturfag. In M. Ødegaard, B. S. Haug, S. M. Mork, & G. O. Sørvik (Eds.), *På forskerføtter i naturfag* (pp. 92–115). Universitetsforlaget.

Mouton, M., & Archer, E. (2019). Legitimation code theory to facilitate transition from high school to first-year biology. *Journal of Biological Education, 53*(1), 2–20. https://doi.org/10.1080/00219266.2017.1420681

Musharrat, T. (2020). Teachers' perceptions about use and challenges of hands-On activities in secondary science classrooms. *European Journal of Education Studies*, *7*(12). https://doi.org/10.46827/ejes.v7i12.3384

O'Hallaron, C. L., Palincsar, A. S., & Schleppegrell, M. J. (2015). Reading science: Using systemic functional linguistics to support critical language awareness. *Linguistics and Education, 32.* https://doi.org/10.1016/j.linged.2015.02.002

Painter, C. (1996). The development of language as a resource for thinking: A linguistic view of learning. In R. Hasan & G. Williams (Eds.), *Literacy in society* (pp. 50–85). Longman.

Polias, J. (2016). *Apprenticing students into science—Doing, talking & writing scientifically.* Lexis Education.

Rose, D. (2011). Beating educational inequality with an integrated reading pedagogy. In F. Christie & A. Simpson (Eds.), *Literacy and social responsibility: Multiple perspectives* (pp. 101–115). Equinox.

Schjøtt, S. R., Lohse, D. M., & Meidell Sigsgaard, A. V. (2024). Language and middle-school science: What do children and teachers gain by implementing semantic waves in teaching? M. Sardag, G. Kaya & M. Sogut (Eds.), *Proceedings book series-I of the ESERA 2023 conference: Connecting science education with cultural heritage*, Strand 18, (pp. 560–575). Nobel Bilimsel Eserler.

Schleppegrell, M. J. (2004). *The language of schooling: A functional linguistics perspective.* Erlbaum.

Windschitl, M., Thompson, J. J., & Braaten, M. L. (2018). *Ambitious science teaching.* Harvard University Press.

Chapter 6
Making the Most of Formative Assessment in Primary Science

Sarah Earle and Regula Grob

Abstract In this chapter, we describe and exemplify formative assessment in primary science as a cyclical process. We identify points at which the cycle can become stalled, explaining why the use of assessment strategies may not always have their desired impact. We also explore opportunities for the formative assessment cycle to be successful within primary science lessons using examples from classrooms in Switzerland and the United Kingdom (UK). From Switzerland, we draw on results from the ASSIST-ME (Assessment In Science, Technology and Mathematics Education) project which explored formative assessment practices in science inquiry classrooms (Grob et al., 2021). From the UK, we draw upon evidence from the TAPS (Teacher Assessment in Primary Science) project and the Focus4TAPS professional learning programme which have been found to have a positive impact on both children's learning and teacher confidence in science teaching (Mujtaba et al., 2022). We argue that formative assessment is essential in primary science, to support children to move from common-sense explanations of the world to more scientific viewpoints. Use of such primary science teacher assessment capability is a developmental process that needs to be supported by ongoing professional learning opportunities and exemplification, enabling the adaptation of lessons to meet the needs of the learners.

Keywords Formative assessment · Primary science · ASSIST-ME · TAPS

S. Earle (✉)
Bath Spa University, Bath, UK
e-mail: s.earle@bathspa.ac.uk

R. Grob
University of Teacher Education Lucerne, Lucerne, Switzerland
e-mail: regula.grob@phlu.ch

S. Earle et al. (eds.), *Primary Science Learning for Children, Teachers, and Communities*, https://doi.org/10.1007/978-981-95-0561-6_6

6.1 Introduction

Formative use of assessment has long been championed as a means of supporting learning (Hattie, 2009), but enacting this in a primary science classroom can be tricky (Gomez & Jakobssen, 2014; Gotwals et al., 2015; Furtak et al., 2008). A range of factors increase the difficulty of formative action in primary science. For example, primary school teachers are often generalists, pedagogy specialists rather than subject specialists, and so may lack confidence with teaching science. Effective use of formative assessment requires a wide range of subject and pedagogical content knowledge (PCK, Shulman, 1986). Limited lesson time for primary science in many contexts, makes it hard to find the time for learners to reflect or act upon any feedback. The practical nature of science also means that it might be more difficult to keep track of learning. In addition, young children are more likely to be talking about their learning, rather than writing up scientific reports, making it challenging to hear from everyone in a busy classroom. Nevertheless, the use of formative assessment is a challenge worth pursuing, because children arrive in primary science classrooms with their own common-sense explanations of the world that need to be elicited and explored, to build scientific knowledge and understanding.

In this chapter, we aim to unpack the elements of formative assessment to support children's learning in the primary science context. We explore the formative assessment process within primary science lessons using examples from classrooms in Switzerland from the *Assessment In Science, Technology and Mathematics Education* (ASSIST-ME) project and in England from the *Teacher Assessment in Primary Science* (TAPS) project. We describe and exemplify formative assessment as a cyclical process and identify points at which the cycle can become stalled and opportunities for it to be successful.

6.1.1 *Switzerland Context and the ASSIST-ME Project*

Science education at primary school level in Switzerland takes place in the context of an integrated subject that involves both the natural and the social sciences. The topics covered are looked at from perspectives of biology, physics, technology, geography, history, sociology, economy and ethics. The curricular guidelines define competences to be covered, but they leave a high degree of freedom to the teachers in deciding on the amount of time spent on particular competences or on deciding on what concrete example a competence should be developed in their science classes. This teaching autonomy, along with the absence of standardised high-stakes summative exams, has a long tradition in Swiss classrooms. Consequently, the teaching practices vary a lot between teachers. Ten years ago, when a new curriculum for all the German-speaking states was introduced, new textbooks for science at primary school level were developed as well. Most teachers had previously developed their own teaching materials but since the emergence of textbooks, more and more teachers, at least partially, include these in their science teaching.

The ASSIST-ME (Assessment In Science, Technology and Mathematics Education) project explored formative assessment practices in science inquiry classrooms. In Switzerland, 20 science teachers explored different methods of formal formative assessment (such as written teacher feedback; peer-assessment; reflective self-assessment) in their inquiry classrooms. The respective insights into the teachers' planning choices, teaching materials and teachers' reflections on their units (typically, a topical unit consists of two to10 school hours) have helped to understand some of the benefits and challenges of formative assessment in practice. Furthermore, a clearer view on potential measures of support for the implementation of formative assessment was gained (Grob et al., 2017, 2021).

Such an explorative study was meaningful for the educational context of Switzerland, where teachers have a lot of freedom in their science teaching. There is a curriculum which defines fundamental competences to be acquired (such as experimentation), but the teachers decide on what topics to work on and how much time to spend on those. The science textbooks are closely linked to the curriculum, but the teachers decide whether they use these textbooks or whether they prefer to use different materials or develop their own programmes. This freedom has a long tradition in the Swiss classrooms, and it is supported by the fact that there are no large-scale high-stake summative tests in science which are evaluated on the level of the individual student (Labudde, 2007).

6.1.2 UK Context and the TAPS Project

In the UK, each of the four nations (England, Scotland, Wales and Northern Ireland) has its own national curriculum requirements and assessment system. In Scotland and England, the science curriculum is based around biology, chemistry and physics, with detailed subject knowledge objectives listed for coverage across the primary years. In Wales, science is combined with technology with broad descriptors for this area of learning, whilst in Northern Ireland science is linked with history and geography within the 'World Around Us' strand. Rather than textbooks, a wide range of subject association resources and commercial schemes of work are available to provide guidance for the teaching of science. Each primary school chooses its own way of implementing the national curriculum, although not necessarily prioritising science. Some school practices are decided at a level above the individual school (e.g. as a multi-academy trust in England or as a school cluster in the other nations), thus there is a wide variety of practice across the sector.

The high stakes accountability context of primary schools in England, with school league tables and a powerful inspectorate, mean that 'assessment' is often driven by a need for summative, trackable numerical data. Whilst primary school standardised national tests for science were actually removed in 2009, science remains a core subject, with teachers needing to make their own assessment judgements for whether children have 'met' the expected standard laid out in the National Curriculum at ages 7 and 11. However, science data is not used in school league tables, meaning that primary science lacks status (and often time) in comparison to the highly scrutinised subjects of English and mathematics. There are a large number of science and STEM organisations in the UK, supporting practice and calling for more attention for science as a means of securing the 'scientist-pipeline' and helping young people to see that science is 'for them', boosting their 'science capital' (Nag Chowdhuri et al., 2021). This complex context leads to a wide variability in terms of practice in primary science, with many schools searching for pre-made schemes of work with lessons that will fit into a crowded timetable. However, the strange situation of being a core subject without the accountability data, also provides an opportunity for formative assessment to take more of a lead, as long as teachers have enough support to develop their PCK.

The TAPS (Teacher Assessment in Primary Science) project began in 2013 in response to a lack of statutory guidance for schools in England regarding how assessment judgements should be made under the new National Curriculum (DfE, 2013). TAPS used a design-based research approach (Anderson & Shattuck, 2012; Davies et al., 2017) to work with schools, collaboratively creating resources to support primary science teaching and assessment. Key products from the project include a bank of focused assessment lessons and children's work examples (see TAPS website below), together with a pyramid-shaped model of 'formative to summative' assessment, whereby information from normal classroom activities can be used formatively to support learning and may be summarised for summative purposes when needed (e.g. to pass information on to next class teacher). The overarching aim at the beginning of the project was to support teacher judgements to be more valid and reliable, through the use of a wider range of information (than just a standalone test) and building a shared understanding of what 'meeting expectations' looked like under the new curriculum. Over time, it became clear that building a shared understanding of progression in primary science was beneficial for formative assessment and it was this in-class responsive teaching where the most impact could be made.

6.2 Formative Assessment Practice

Formative assessment is focused on supporting the learner. It may involve observing, listening and eliciting children's ideas, with this feedback from learners utilised by the teacher or the learners to inform decisions about what to do next. Formative assessment within class teaching may be difficult to identify, because it is an embedded part of the teaching and learning process (Wiliam, 2010). The popular term 'Assessment for Learning' (AfL) is also often used to describe formative assessment since it emphasises the importance of supporting learning as part of classroom practice. The UK-based Assessment Reform Group defined AfL as: *"the process of seeking and interpreting evidence for use by learners and their teachers to decide where the learners are in their learning, where they need to go and how best to get there"*

(ARG, 2002, p.2). A key point here is that it is a process, the assessment task itself is just one part of this process, there is a need to act upon the information. Some more recent definitions emphasise this notion of 'action', claiming that assessment can only be formative if the information gathered through assessment is used to advance further learning. For example:

> *An assessment functions formatively to the extent that evidence about student achievement is elicited, interpreted, and used by teachers, learners and their peers to make decisions about the next steps in instruction that are likely to be better, or better founded, than the decisions they would have made in the absence of that evidence.* (Wiliam, 2017, p.43)

Formative assessment can be described as cyclical. Harlen (2013: 18) represented the cycle as moving from a student activity, to collection of evidence, making of judgements and decisions regarding next steps for the following student activity. During our work with primary school teachers and academics, through the projects described above, we found that we wanted to tweak the wording in the cycle slightly, for example, to explicitly talk about 'children's learning' rather than 'evidence' from 'students'. In Fig. 6.1, we present our own version of this cycle, making explicit the processes (arrows) and outcomes (boxes at each stage). Following children's engagement in an activity, 'data collection' could simply be teacher observations or it could be the children recording their learning in some way, for example, by drawing, model-making or writing. When the teacher interprets this 'data' or information about children's learning, they can then use their judgements for summative assessment purposes or for a decision on next steps of learning. In the latter case, formative use of the information on children's learning could lead to a new activity being planned and enacted, which starts off a second round in the formative assessment cycle. In Dolin et al.'s (2018) cycle, they place 'students' in the centre of the cycle to emphasise their participation at all stages. We also feel that children should be involved in the whole process, but see them as an embedded part of the cycle, essentially following around with each stage, particularly if this was applied to self- or peer-assessment in class. We decided to place 'learning goals and criteria' in the centre of the cycle, because emphasis on this at each stage can be helpful to keep the focus on learning, as teachers (or children in the case of self- or peer-assessment) consider what they are trying to achieve in the lesson and whether they are demonstrating meeting of the lesson goal(s).

6.2.1 Two Primary Science Examples of the Formative Assessment Cycle

To exemplify the formative assessment cycle detailed above in the context of primary science and design technology, two lesson examples are presented below.

Example 1: Shadows in a science lesson in a class of 7 and 8 year olds in England

Learning goal: Consider how a shadow is made.

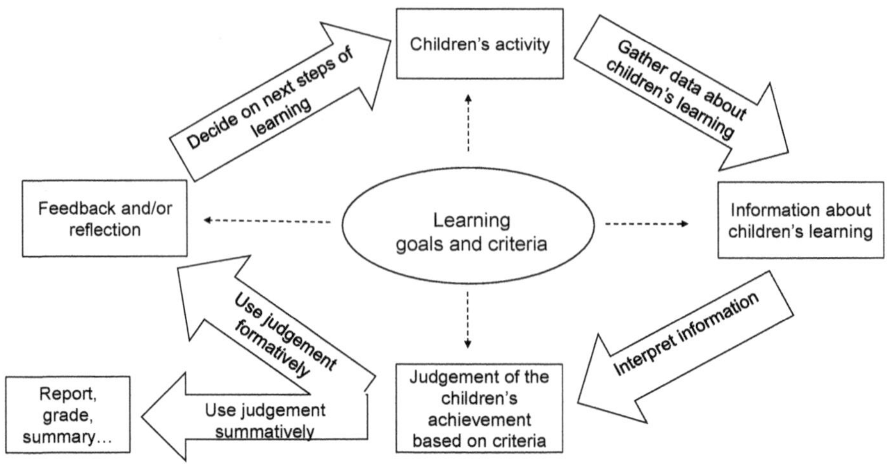

Fig. 6.1 Formative assessment cycle (adapted from Harlen, 2013). Arrows refer to processes, boxes refer to results of the aforementioned processes

Children's activity: Look at the shadows made by their own bodies in the playground. Draw and label their ideas about how a shadow is made.

Information about children's learning: A range of ideas are present in the children's drawings, including the sun as a light source, the child themselves and their shadow on the floor. Their labelling and writing varies in how much scientific vocabulary is used.

Judgement: Most children show a light source and a shadow of the right shape, but not all have the shadow in the right position (opposite the light source). Many struggled to explain and some used the wrong scientific words e.g. the shadow 'reflects' rather than the opaque object blocks the light.

Feedback/next step: Teacher recognises that they need to model using scientific vocabulary in discussion with the children over the next lessons.

Next activity: Ask the children to show how to make shadows with a torch and opaque object, recording with new labelled diagrams (Fig. 6.2).

Example 2: Constructing and deconstructing a clock in a Swiss class of 8 year olds

Learning goal: Construct and deconstruct a clock in order to answer the question: How does a clock function?

Children's activity: Develop more specific questions (e.g. How do the hands move? What is the blue toothed wheel important for? How can the time be adjusted manually?) and try to answer them by constructing and deconstructing a clock from a construction kit; present all stages of the inquiry process in whole-class discussions in order to provide the teacher (and the other students) with the opportunity to provide feedback and help with refined questions.

Information about children's learning: An overview of the children's level of understanding can be gained in the whole-class discussions. There, the teacher has

Fig. 6.2 Creating a shadow

the opportunity to assess the childrens' technical understanding on-the-fly and to provide feedback on the next steps in the inquiry process.

Judgement: Often, the students have difficulties in systematically trying to answer a question—they are interested in tinkering with the clock but not always in a purposeful way.

Feedback/Reflection: The teacher's feedback is often embedded in the ongoing whole-class discussions. There, the teacher moderates the discussion in a way that the students can refocus on the question or, if properly solved, revise the question.

Next activity: Amend the construction/deconstruction process based on the original or on a revised question (Fig. 6.3).

6.2.2 Beyond a Simple Formative Assessment Cycle

The formative assessment cycle in Fig. 6.1 and subsequent examples of practice present a simplistic picture of what could happen in a learning sequence. In reality however, the picture is much messier. For instance, there could be many mini-cycles within lesson adaptations, in the moment or 'on-the-fly' tweaks to instructions, activities or groupings where the skilled practitioner is constantly responding to the needs of their class (e.g., Ruiz-Primo & Furtak, 2004, 2006; Shavelson et al., 2008). The generic cycle does not specifically mention the teacher, enabling self- and peer-assessment to be considered, for example, with peer feedback providing the opportunity for reflection in the next student activity (e.g., Andrade, 2010; Topping, 2003). The cycle also represents a whole class view, not explicitly considering how different individuals may need different adaptations at different points in the lesson or series of lessons. Again, the skilled practitioner will be tweaking the children's activities and instructions to ensure accessibility and appropriate challenge for learners.

Nevertheless, even discounting within lesson and individual adaptations, it is proposed that this simplistic cycle can be hard to replicate in the primary science classroom, because in reality there are many potential breaks or interruptions in the cycle. For instance, assessment information in the form of evidence from children's

Fig. 6.3 A Clock constructed from a kit

activities can be generated, but then not used formatively for a range of reasons. Formative use may not happen, for example, due to time constraints for the teaching of science, lack of teacher knowledge about progression and next steps, a rigid curriculum with set lessons across the term or a focus on non-science aspects of learning such as handwriting. Different potential breaks in the formative assessment cycle are presented in Table 6.1, with primary science classroom examples constructed from the authors' experience in the two projects.

Consequently, it is only logical to next discuss why formative assessment is hard to accomplish in the classroom. In-the-moment adaptations and responsive teaching, as well as more formal, preplanned formative assessment methods, all require a lot of knowledge from practitioners (Ateh, 2015; Gotwals et al., 2015). Teachers need to know about the content they are trying to teach, suitable pedagogical approaches, possible misconceptions and likely progression (Bennett, 2011; Furtak et al., 2008, 2016; Yin et al., 2008); otherwise known as pedagogical content knowledge (PCK) as termed by Shulman (1986). Discipline-specific examples, for science in this instance, are needed to enable teachers to translate generic principles and strategies for the particular topic they are trying to teach (Gotwals & Cisterna, 2022; Heredia, 2020; Quinlan & Pitt, 2021). Teachers also need to be aware of the unique context of the children in their class, their previous learning experiences and their current needs. This is a lot to think about when you are trying to do practical science activities with 30 children. It is therefore not surprising that formative use of primary science

Table 6.1 Potential breaks in the formative assessment cycle

Part of the cycle	Potential break in the cycle	Examples of what this might look like in the primary science classroom
Children's activity	Children's activities can be selected without science learning in mind. The focus may be on the activity rather than the science learning.	*Making an edible model sounded fun. I'm not sure how much science they got out of it but they all enjoyed it!*
Information about children's learning	Children complete the activity and record their learning. The lesson ends and the next lesson moves on to another topic. The focus is on completing the activity. There is no time to analyse children's written work between lessons.	*Drawing and writing about the activity takes the rest of the lesson and we need to move on, so a quick check that they have written neatly is all I have time for this week.*
Teacher judgement	Sometimes teachers might not know what it looks like to achieve the lesson goal. Judgements may be recorded for summative purposes, but not used to support next steps in learning. The focus may be on recording a summative judgement, without the diagnostic detail to help decide on next steps.	*I followed the lesson plan and they did what was asked, so I assume this objective is covered.* *I sorted the learners' outcomes into 'correct', 'nearly there' and 'wrong', then record this on the summative assessment list. For the ones who have not yet 'got it', maybe they will come back to this topic next year.* The judgement does not include details on what is missing for a correct / "good" solution but just contains general statements from which it is hard to learn.
Feedback	Feedback is given to students, but they do not have an opportunity to act upon it.	*The children's books include 'next step' prompts, but there is no time to discuss or come back to this.* *Feedback on relevant competencies is provided, but the subsequent activity requires different competencies.*
Next step	A gap in children's learning is identified diagnostically, but the next activity/lesson is not changed because of time, uncertainty about what to do next or a rigid curriculum plan.	*I know there are some that haven't really got it, but there's a lot more to cover before the end of term, so I can't hold everyone else up. We'll come back to it if/when we can.*

assessment is hard to accomplish in practice. To come 'off plan' and adapt the lesson in response to the children's replies (and to feel that you are allowed to do this if following a particular scheme of work or textbook), requires expertise and confidence, sometimes referred to as 'AfL capability' (Smaill, 2020).

Contrary to on-the-fly formative assessment, elicitation activities, such as annotated drawings, multiple choice quiz questions or 'odd one out' discussions, can be planned in advance. International results show that such formal or pre planned methods can be used meaningfully by teachers, particularly when introduced in continuous professional development and when it can be based on transferrable materials and examples (Hondrich et al., 2016).

In order to further explore formative use of assessment in primary science practice, the next section will consider examples of success from projects in Switzerland and England. The analysis will be the basis for our conclusions and recommendations on how to make effective use of the formative assessment cycle in the primary science classroom.

6.3 Stories of Practice from the ASSIST-ME Project

As noted above, the freedom in Swiss classrooms provides ample room for the teachers to shape their teaching according to the needs of their children. In this story of practice, an example of successful formative assessment from a Swiss science classroom will be explored. Emphasis will be put on the question of what made the example a success or, in other words, how failure could be avoided.

A teacher at 3rd grade primary school level (aged 9 and 10 years) let her class observe the growth of chicks (young chickens). She had a group of chicks near her classroom and the children observed them every morning for 5–10 min. For this observation task, the children had to think of specific questions to focus their observations on these questions, and they also had to individually write a short paragraph in the so-called 'chick journal'. The teacher collected the chick journals every day and briefly commented on the latest entries (see Fig. 6.4). She focussed her written feedback on (a) the distinction between observations and claims/conclusions and (b) on the precision of their descriptions and sketches. The children were often told to read the teacher's comments and to check for improvement but also to read the journal entries of their peers. In addition to this, the teacher planned and conducted several short inputs and group discussions on good observations and on the distinction between observations and claims throughout the unit.

In the evaluation form, the teacher reported that 'pointing at these two assessment criteria over a long period of time really helped the students to keep focused'. She said in the interview that 'it was easy to remind both the students but also their parents of the goals of that unit'. The teacher also realised that 'it took some time before the students really understood the assessment criteria but also the principle of this type of formative feedback. Some of the students really needed the external encouragement to engage with the feedback and to plan next steps in learning'. She also mentioned that 'the extra workload for providing feedback to every student every day is only worth the effort when the assessment criteria are very relevant in science education'. So, she recommended 'spending ample time of the preparation phase on the selection and formulation of suitable criteria'.

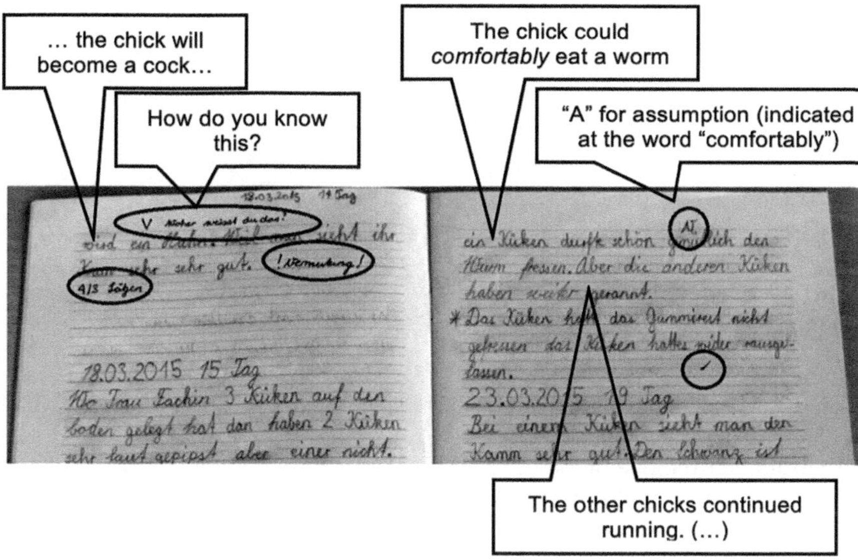

Fig. 6.4 Example of a student's chick journal with the teacher's comments (teacher's comments circled in black)

In the group discussion where the teacher shared her experiences with this trial, her teacher colleagues were immediately convinced about the effect of the many rounds of individual feedback and the many subsequent opportunities for the children to engage with this feedback. However, they shared the teacher's impression that such a time-consuming, demanding setting needs careful placement in the semester and thoughtful selection of the assessment criteria in the way of 'which competence is so important that a lot of preparation and lesson time should be spent on it'.

6.4 Stories of Practice from the TAPS Project

Out of the TAPS project, a professional development (PD) programme called Focus4TAPS was developed, to help teachers to implement the approach, with particular assistance for the teaching and assessment of science inquiry. The Focus4TAPS programme has been externally evaluated in a large-scale randomised control trial and found to have a positive impact on both children's learning and teacher confidence (Mujtaba et al., 2022). The programme involves three PD days, with teachers bringing examples from class to share on the second and third days, to support discussions about progression and next steps. A normal next step for primary science classrooms in England would be to move on to the next lesson or topic, because of the crowded timetable, but the PD promotes the formative use of assessment information (Earle,

Fig. 6.5 Sugar cube stack in coloured water

2021), to continue the path on the formative assessment cycle, rather than move off and on to the next new topic.

An example from one of the PD days involves a discussion of a project school's classroom activity with the children's outcomes for creating their own results tables. Teachers are invited to experience the activity within the PD, so that they can consider how something similar would work in their classrooms. They place a stack of three sugar cubes into a small pool of coloured water and observe (Fig. 6.5). The coloured water begins to rise up the sugar cubes and it looks like the colour will go to the top, but then the cubes fall down, because the bottom cube has begun to dissolve. This can lead to discussions around dissolving and which variables might affect the time taken for the stack to fall over, for example, amount of water or cubes, temperature of water, type or amount of food colouring, or items placed between the cubes such as paper or foil. Participants are asked to do a brief investigation into one of these variables and record their results. After discussing their findings, a story of practice is shared in the PD from a teacher who we will call Emily.

Emily followed the TAPS example lesson plan to do the sugar cubes lesson with her class of 9 and 10 year olds. She was surprised at the way the children had recorded their results. Some had made lists and some had attempted to draw their own results table, but most were unclear, for example, missing column headings so that the meaning of the numbers was uncertain. Initially Emily felt that her lesson had not worked as it was meant to and felt disappointed. However, after discussion with colleagues, she realised that the lesson had shown her something that the children struggled with, a gap in their learning. This led Emily to model how to draw a results table and give the children an opportunity to apply this in a future inquiry, where to Emily's relief, they were able to present their results much more clearly. This learning story demonstrates the formative use of assessment, with changes to teaching made in response to feedback from the children. By honing in on one part of the inquiry in the focused assessment lesson, it meant that a learning gap was identified and addressed, the formative assessment cycle was completed.

6.5 Conclusion: Making the Most of Formative Assessment

Through exploration of stories of practice from primary science, we hope to have made the case for use of formative assessment to support children's learning. We expect most readers may already have recognised the large evidence base for formative use of assessment, however, the reality of making this work in the classroom has been more elusive. The formative assessment cycle often starts off well, with planned elicitation strategies to find out about children's scientific ideas, or purposeful activities designed to support scientific learning. But, as noted in Table 6.1, there are many places where the cycle may stall, where effective use is not made of formative assessment information and an opportunity for learning is lost. Taking a more positive view, rather than list points in the cycle where things might go wrong, let us utilise our stories of practice from above to highlight opportunities for where the formative assessment cycle can work well.

As demonstrated in Table 6.2, opportunities for success are available throughout the formative assessment cycle. However, discipline-specific teacher assessment capability takes time to build, needing professional learning opportunities for the development of PCK that enables effective implementation. Completing formative assessment cycles is worth the effort, because it can efficiently address learner needs, providing visible impact on science attainment. In addition, the teacher knowledge that comes with the development of formative assessment capability, with recognition of learning goals, noticing and interpreting, is also applicable across the curriculum.

The learning stories from the UK and Switzerland discussed above show the potential of the formative use of assessment to support children's learning. The examples and discussion above also demonstrate that formative assessment takes place in a particular educational context, which extends beyond the responsibility of the class and their teacher. Consequently, successful implementation of formative assessment does not only depend on the individual teacher, but could be supported by other stakeholders, including:

- **Children:** Explicit discussion with children about formative use of assessment can support them to understand and take ownership of their own learning processes, 'activating' them as learning resources for each other via self and peer assessment (Wiliam, 2017).
- **Science leads and school leaders:** As discussed in Table 6.2, the judgement part of the formative assessment cycle may become more successful when teachers are given the opportunity to discuss their children's learning and their own judgements in a group. Providing opportunities for teachers from different age groups to discuss attainment can support the development of a shared understanding of progression in primary science. The establishment of such communities of practice could be supported and facilitated by teaching teams and school leaders.
- **Teacher educators:** As discussed in Tables 6.1 and 6.2, it is not easy for teachers to interpret information on children's learning and to use this information to plan subsequent steps in teaching and learning. Teacher educators could facilitate

Table 6.2 Potential successes in the formative assessment cycle

Part of the cycle	Potential successes in the cycle	Classroom examples
Children's activity	Choose an initial activity to match the learning goal(s). Being able to adapt subsequent activities to meet the needs of the learners for further formative assessment cycles. Activity choice may be informed by elicitation of children's pre-existing ideas.	*To develop understanding of shadows, torches were used to make shadows of different shapes and sizes.* *To develop close observations over time, the growth of chicks or plants were studied.* *To explore construction, clock mechanisms were investigated.* *Starting a unit of work with a discussion, to find out children's current ideas, for example, annotated drawings or an odd-one-out.*
Information about children's learning	Children are asked to talk, draw or write about their learning. Their recording focuses on the particular science learning goal, rather than making sure everyone has written everything down in overly long reports.	*When observation over time was the focus, a chick journal was kept.* *When recording results was the focus, a table of results was completed.* *When understanding of shadows was the focus, annotated diagrams were drawn.*
Judgement	Children's outcomes are considered in relation to the science learning goals. Teachers discuss outcomes with other teachers and/or look at examples of criteria/work appropriate to curriculum context. This may include looking at objectives/examples for older/younger children, to support progression Judgements are used to inform feedback and decisions regarding next steps in learning.	*Diagnostic judgements about shadows understanding helped to spot a gap in scientific vocabulary use.* *Explicit success criteria for the science learning goal are discussed, to enable self and peer assessment within the lesson.* *Teachers talk to each other about the science learning in their class, to consider what meeting the learning goals looks like in practice.*
Feedback and next steps	Time is planned for consideration of teacher/peer feedback or for self-reflection. Feedback is specific to the learning goals. There is the opportunity to act on feedback (this could be within the same lesson or in a subsequent lesson). Next activities are planned to build on what came before.	*After attempting to make their own branching keys with yes/no questions to identify the animals, children test out each other's and give feedback to improve.* *A judgement that children struggled drawing results tables, meant that the teacher modelled results tables in the next lesson.*

these steps in the formative assessment cycle in initial and continuous professional development by explaining its importance theoretically and with examples.

- **Resource developers:** One conclusion from the examples discussed in the chapters is that it can be hard to operationalize goals from the science curriculum to concrete activities, learning goals or respective criteria of success. Resource developers could aid this with examples of science competences that are operationalized to a degree they can be used in the classroom.
- **Researchers:** As the overall impression of this chapter reflects, there still is a wide research-practice gap in terms of formative assessment in the science classroom. Whilst some questions have been worked on in research, many findings have not yet been incorporated into daily science teaching practice. Working collaboratively with teachers and schools can provide opportunities for close-to-practice research, enabling more applicable solutions and real classroom examples to be shared.
- **Policy makers:** The role of formative assessment could be strengthened in official documents, so that formative assessment is perceived as something essential to good science teaching and learning rather than a nice-to-have add-on.

Each context also has its own challenges, with professional freedom in Switzerland placing a burden of choice on the teacher, whilst the opposite arises in England, with teachers needing support to make changes to prescribed learning programmes to adapt lessons to meet the needs of the learners in their class. Use of formative assessment is particularly important in primary science, because children come to this subject with their own common-sense explanations for how the world works, which may not match the scientific view. By eliciting these ideas, and then acting upon that information to adapt future learning opportunities, children's science learning can be enhanced.

References

Anderson, T., & Shattuck, J. (2012). Design-based research: A decade of progress in education research? *Educational Researcher, 41*(1), 16–25.

Andrade, H. (2010). Students as the definitive source of formative assessment. In H. Andrade & G. J. Cizek (Eds.), *Handbook of formative assessment* (pp. 90–105). Routledge.

Assessment Reform Group (ARG). (2002). *Assessment for learning: 10 Principles: Research-based principles to guide classroom practice.* University of Cambridge School of Education.

Ateh, C. (2015). Science teachers' elicitation practices: Insights for formative assessment. *Educational Assessment, 20*(12), 112–131.

Bennett, R. E. (2011). Formative assessment: A critical review. *Assessment in Education: Principles, Policy & Practice, 18*(1), 5–25.

Davies, D., Earle, S., McMahon, K., Howe, A., & Collier, C. (2017). Development and exemplification of a model for teacher assessment in primary science. *International Journal of Science Education, 39*(14), 1869–1890.

Department for Education (DfE). (2013). *National curriculum in England: Science programmes of study*. DfE.

Dolin, J., Black, P., Harlen, W., & Tiberghien, A. (2018). Exploring relations between formative and summative assessment. In J. Dolin & R. Evans (Éd.), *Transforming assessment* (Vol. 4, pp. 53–80). Springer International Publishing. https://doi.org/10.1007/978-3-319-63248-3

Earle, S. (2021). Formative decision-making in response to primary science classroom assessment: What to do next? *Frontiers in Education, 5*, Article 584200.

Furtak, E. M., Ruiz-Primo, M. A., Shemwell, J. T., Ayala, C. C., Brandon, P. R., Shavelson, R. J., & Yin, Y. (2008). On the fidelity of implementing embedded formative assessments and its relation to student learning. *Applied Measurement in Education, 21*(4), 360–389.

Furtak, E. M., Keimer, K., Circi, R. K., Swanson, R., deLeon, V., Morrison, D., & Heredia, S. (2016). Teachers' formative assessment abilities and their relationship to student learning: Findings from a four-year intervention study. *Instructional Science, 44*, 267–291.

Gomez, M., & Jakobsson, A. (2014). Everyday classroom assessment practices in science classrooms in Sweden. *Cultural Studies of Science Education, 9*(4), 825–853. https://doi.org/10.1007/s11422-014-9595-y

Gotwals, A. W., Philhower, J., Cisterna, D., & Bennett, S. (2015). Using video to examine formative assessment practices as a measure of expertise for mathematics and science teachers. [Special Issue]. *International Journal of Science and Mathematics Education, 13*(2), 405–423.

Gotwals, A. W., & Cisterna, D. (2022). Formative assessment practice progressions for teacher preparation: A framework and illustrative case. *Teacher and Teacher Education, 110*, 1–13.

Grob, R., Holmeier, M., & Labudde, P. (2017). Formative assessment to support students' competences in inquiry-based science education. *Interdisciplinary Journal of Problem-Based Learning., 11*(2), 1–8.

Grob, R., Holmeier, M., & Labudde, P. (2021). Analysing formal formative assessment activities in the context of inquiry at primary and upper secondary school in Switzerland. *International Journal of Science Education, 43*(3), 407–427.

Harlen, W. (2013). *Assessment and inquiry-based science education: Issues in policy and practice.* Inter Academy Partnership.

Hattie, J. (2009). *Visible learning. A synthesis of over 800 meta-analyses relating to achievement.* Routledge.

Heredia, S. C. (2020). Exploring the role of coherence in science teachers' sense- making of science-specific formative assessment in professional development. *Science Education, 104*(3), 581–604.

Hondrich, A. L., Hertel, S., Adl-Amini, K., & Klieme, E. (2016). Implementing curriculum-embedded formative assessment in primary school science classrooms. *Assessment in Education: Principles, Policy & Practice, 23*(3), 353–376.

Labudde, P. (2007). How to develop, implement and assess standards in science education? 12 challenges from a Swiss perspective. In D. Waddington, P. Nentwig, & S. Schanze (Eds.), *Making it possible: Standards in science education* (pp. 277–301). Waxmann.

Mujtaba, T., Sheldrake, R., Hodgen, J., & Reiss, M. (2022). *Focus for teacher assessment of primary science (Focus4TAPS): Evaluation report.* Education Endowment Foundation. Retrieved April 23, 2024, from https://educationendowmentfoundation.org.uk/projects-and-evaluation/projects/focus4taps

Nag Chowdhuri, M., King, H., & Archer, L. (2021). *The primary science capital teaching approach: Teacher handbook.* UCL Institute of Education.

Quinlan, K., & Pitt, E. (2021). Towards signature assessment and feedback practices: A taxonomy of discipline-specific elements of assessment for learning. *Assessment in Education: Principles, Policy & Practice, 28*(2), 191–207.

Ruiz-Primo, M. A., & Furtak, E. M. (2004). *Informal formative assessment of students' understanding of scientific inquiry. CSE Report 639.* University of California.

Ruiz-Primo, M. A., & Furtak, E. M. (2006). Informal formative assessment and scientific inquiry: Exploring teachers' practices and student learning. *Educational Assessment, 11*(3 & 4), 205–235.

Shavelson, R. L., Young, D. B., Ayala, C. C., Brandon, P. R., Furtak, E. M., Ruiz-Primo, M. A., Tomita, M. K., & Yin, Y. (2008). On the impact of curriculum-embedded formative assessment on learning: A collaboration between curriculum and assessment developers. *Applied Measurement in Education, 21*, 295–314.

Shulman, L. (1986). Those who understand: Knowledge growth in teaching. *Educational Researcher, 15*(2), 4–14.

Smaill, E. (2020). Using involvement in moderation to strengthen teachers' assessment for learning capability. *Assessment in Education: Principles, Policy & Practice, 27*(5), 522–543.

Teacher Assessment in Primary Science (TAPS) website. Retrieved February 23, 2024, from https://pstt.org.uk/unique-resources/taps/

Topping, K. J. (2003). Self and peer assessment in school and university: Reliability, validity and utility. In M. Segers, F. Dochy, & E. Cascaller (Eds.), *Optimising new modes of assessment: In search of qualities and standards* (pp. 55–87). Kluwer Academic Publishers.

Wiliam, D. (2010). Research literature and implications for a new theory of formative assessment. In H. Andrade & G. J. Cizek (Eds.), *Handbook of formative assessment* (pp. 18–40). Routledge.

Wiliam, D. (2017). *Embedded formative assessment* (2nd ed.). Solution Tree Press.

Yin, Y., Shavelson, R. J., Ayala, C. C., Ruiz-Primo, M. A., Brandon, P. R., Furtak, E. M., Tomita, M. K., & Young, D. B. (2008). On the impact of formative assessment on student motivation, achievement, and conceptual change. *Applied Measurement in Education, 21*(4), 335–359.

Part II
Primary Science Learning for Teachers

Chapter 7
Nurturing Tomorrow's STEM Educators: Immersive Experiences for Primary Generalist Pre-service Teachers

Melanie Nash, Ondine Bradbury, and Ange Fitzgerald

Abstract This chapter presents an innovative approach to equipping generalist primary pre-service teachers with essential competencies and enthusiasm for effective STEM education. It explores the transformative experiences of a postgraduate-level cohort of pre-service teachers who participated in an immersive STEM course as part of their initial teacher education program. What distinguishes this study is the diverse backgrounds of the pre-service teachers, many of whom are 'career changers' bringing rich experiences from various fields. The primary focus of this research is to examine how the immersive STEM course influenced the learning and teaching perspectives of these pre-service teachers, as well as the educator who designed and facilitated the course. Practice architectures provided the framework for analysing the data by offering insights into how the course design and interactions within the learning environment influenced participant outcomes. To gather these insights, a combination of research methods was employed, including surveys and focus group interviews. Surveys captured information on pre-service teachers' confidence and experience from the STEM course, while focus group interviews explored their reflections, and the challenges encountered during the course. Findings are presented through narrative vignettes that feature the voices of the pre-service teachers and the educator involved in course design and facilitation.

Keyword STEM education · Initial teacher education · Immersive experiences · Pre-service primary teachers

M. Nash (✉) · A. Fitzgerald
RMIT University, Melbourne, Australia
e-mail: melanie.nash@rmit.edu.au

O. Bradbury
Swinburne University of Technology, Melbourne, Australia

S. Earle et al. (eds.), *Primary Science Learning for Children, Teachers, and Communities*, https://doi.org/10.1007/978-981-95-0561-6_7

7.1 Introduction

This chapter provides a nuanced exploration of how a cohort of career-changing master's teachers navigated an immersive STEM course, offering valuable insights into the preparation of future STEM educators. It advocates for the systematic integration of immersive STEM experiences into teacher preparation programs, emphasising the continuous support needed for future educators to excel in STEM education.

This chapter examined the impact of the compulsory STEM course 'Extending STEM' offered in a postgraduate pre-service teacher program at RMIT University in Melbourne, Australia. The course was designed for postgraduate pre-service teachers enrolled in a two-year Master of Teaching program preparing them for primary generalist teaching roles. The course endeavoured to deepen the pre-service generalist teachers' understanding of STEM disciplines and cultivate their advocacy for STEM education in classrooms. By challenging preconceived notions and showcasing STEM's transformative potential, the course aimed to equip pre-service generalist teachers to inspire and prepare their students for a STEM-centric future.

The ongoing shortage in the teacher pipeline (Australian Institute for Teaching and School Leadership [AITSL], 2021) has prompted Initial Teacher Education (ITE) programs to adopt increasingly intensive or accelerated course structures (Collins et al., 2013). These courses, are often condensed into brief periods like five or six days, aimed to accelerate learning and streamline the preparation of pre-service teachers for the demands of modern classrooms. However, despite their practicality in delivering essential content efficiently, feedback from pre-service teachers has revealed a significant concern: a lack of enthusiasm for this fast-paced, content-focused approach (Collins et al., 2013).

The 'Extending STEM' course at RMIT was scheduled in such an intensive mode, spread over five weeks, with each week comprising a single day of learning that lasted for approximately five hours. Mindful of the negative response pre-service teachers had expressed for previous intensive courses experienced during their program, the teacher educator wanted to avoid a passive heavily content-focused approach. Consequently, she crafted the course to be an immersive experience focused on an integrated hands-on project-based learning approach, where pre-service teachers were exposed to course design that reflected the principles of andragogy. This approach, as highlighted by Gilstrap (2013), has been shown to make course and program outcomes more relevant to pre-service teachers, thereby enhancing the overall effectiveness of curriculum design.

The course was aligned with several United Nations (UN) Sustainable Development Goals—particularly Goal 11 on sustainable cities and Goal 15 (UNGA, 2015) focusing on biodiversity and ecosystems—the curriculum aimed to equip pre-service teachers with the knowledge and skills to address pressing environmental and educational challenges. The course curriculum emphasised real-world problem-solving and authentic scientific inquiry. For example, pre-service teachers engaged in citizen science projects where they collected data and contributed to ongoing scientific research. Activities included monitoring local biodiversity and assessing environmental impacts and disasters in urban settings, thereby promoting a hands-on understanding of STEM applications in contemporary contexts.

Collaborative partnerships with educational institutions and industry experts enriched the course experience. For example, collaboration with the Banyule Nillumbik Tech School provided insights into design principles and generative AI (Artificial Intelligence), used to create prototypes of sustainable cityscapes. Partnerships with RMIT's geospatial team and Esri a leading geospatial mapping and data analytics company, offered expertise in geospatial technologies, providing tools and resources that enhanced pre-service teachers' technological literacy and practical skills.

The course culminated in a micro-teaching activity at a local primary school, where pre-service teachers led sessions for grade 5/6 students (10–11-year-olds) on disaster risk reduction, biodiversity enhancement, and sustainable water practices. This practical teaching experience aimed to build pre-service teachers' confidence in applying STEM education principles in school settings.

This course 'Extending STEM' was designed to empower pre-service teachers with practical insights into STEM's real-world applications, preparing them to meaningfully engage in educational and environmental challenges throughout their future teaching careers.

7.2 Positioning Within the Literature

7.2.1 STEM Education in Primary Education and ITE Settings

In recent decades, STEM education has taken centre stage in primary and secondary education as governments prioritise the development of a technologically skilled workforce. ITE institutions and educators play a pivotal role in shaping pre-service teachers' understanding of STEM and way STEM courses are included in ITE curricula significantly influences the preparation of pre-service teachers to effectively teach STEM in their future classrooms. By providing a strong foundation in STEM content and pedagogy, ITE institutions can equip pre-service teachers with the knowledge and skills necessary to inspire and engage students in STEM learning.

Additionally, the beliefs and attitudes of teacher educators towards STEM education significantly shape the design and implementation of teacher education programs. This, in turn, influences pre-service teachers' approaches to STEM in their future classrooms. As Pimthong and Williams (2021) suggest, teacher educators who embrace constructivist methodologies and foster interdisciplinary learning are better equipped to prepare pre-service teachers for the challenges of STEM education. Moreover, providing ITE educators with professional development opportunities and fostering connections with STEM experts can enhance their capacity to integrate STEM effectively into their curricula at both primary and secondary levels (Hurley et al., 2024; White et al., 2018).

In primary classroom contexts, many teachers lack confidence in teaching STEM content due to limited backgrounds in these subjects, often leading to rote learning (Sahito & Wassan, 2024) instead of implementing practical and engaging STEM-based approaches to teaching and learning (Kirkiç & Arikan, 2023). This issue is highlighted in McDonald's (2016) review of the contributions of STEM to educational contexts. McDonald (2016) notes that many teachers adopt transmissive pedagogies due to inadequate content knowledge and pedagogical skills, which can influence students' engagement and interest in STEM subjects. Furthermore, this review emphasises the need for professional development to equip teachers with effective STEM pedagogical practices, such as STEM literacy and inquiry-based learning, which can significantly enhance student motivation and achievement. The transition from STEM learning experiences in primary to secondary school is also critical, as it can disrupt students' interest in STEM, highlighting the importance of early and continuous engagement in STEM education. Kirkiç and Arikan (2023) discuss their findings on primary teachers' attitudes toward the teaching of STEM, revealing the competencies associated with teaching STEM are essential. This issue is mirrored in ITE settings, where there is insufficient emphasis on equipping future teachers with necessary STEM skills.

7.2.2 Challenges When Embedding Science and STEM Teaching Experiences

One of the primary challenges in embedding STEM units is the curriculum design itself. The literature indicates that there is a lack of consensus on the definitions and frameworks of STEM, iSTEM, and STEAM, which leads to ambiguity in their implementation (Razi & Zhou, 2022). Educators often find it difficult to integrate these frameworks into a cohesive curriculum that aligns with educational standards and meets the diverse needs of students. Another significant challenge is the availability of resources. Implementing STEM units often requires specialised equipment, materials, and technological tools, which can be costly and difficult to procure. Schools, particularly those in underfunded areas, may find it challenging to provide the necessary resources to support effective STEM education (Razi & Zhou, 2022). Integrating STEM education into the crowded primary curriculum is also a challenge, as time and resources are limited, often resulting in superficial coverage of STEM topics. For many schools, accessibility to STEM resources varies widely.

Addressing these challenges requires policy changes, professional development, and a cultural shift towards valuing STEM education from an early age. According to Lange et al. (2022), implementing a curriculum-agnostic professional learning model has shown significant positive impacts on educators' attitudes, beliefs, confidence, and knowledge in teaching STEM subjects. This approach involves interactive workshops, reflective coaching, and professional learning communities, which collectively help teachers feel more competent and enthusiastic about integrating STEM into their classrooms. The study highlights that professional development tailored to the needs of diverse classrooms, including students from English as an additional language or dialect, is essential for improving teaching practices Lange et al. (2022). By supporting teachers through these multifaceted professional learning experiences, educators can better incorporate developmentally appropriate STEM activities that engage all students, regardless of their linguistic or socio-economic background. This, in turn, promotes a more inclusive and effective STEM education environment, addressing the broader policy and cultural shifts needed to prioritise STEM from an early age.

7.2.3 Generalist Versus Specialists—Teaching Outside of the 'STEM' Field

The debate between generalist and specialist teachers in STEM education highlights the distinct challenges associated with teaching within and outside of a field of expertise. Generalist teachers, often found in primary and lower secondary education, are required to teach a broad range of subjects, including STEM, for which they may not have specialised training. They may require a broad-based teacher education program to equip them with the knowledge and skills to teach various subjects. STEM specialists, however, need focused training in their specific areas to ensure they have the depth of knowledge necessary to teach complex concepts effectively. Research shows that this out-of-field teaching phenomenon can negatively impact the quality of education, resulting in less effective instructional practices, lower student engagement, and diminished achievement in STEM subjects (Du Plessis, 2020; Weldon, 2016). The challenges for generalists may also include maintaining adequate knowledge across all subjects and addressing the specific needs of students in each area. This lack of pedagogical content knowledge specific to STEM subjects can hinder their ability to deliver effective instruction and connect concepts across disciplines.

STEM specialists can cultivate deep interest and expertise in their field, often inspiring students to pursue advanced studies or careers in STEM-related areas. For STEM specialists, the challenge often lies in integrating their specialised knowledge into a broader primary curriculum and collaborating with generalist colleagues to provide a cohesive learning experience. The study by Rinke et al. (2016) highlights that pre-service teachers in a combined STEM methods course reported significantly greater gains in STEM teaching efficacy compared to those in traditional routes. The constraints identified by Rinke et al. (2016) in STEM teacher education, particularly in the subject of Technology, emphasise the need for targeted improvements in the professional development space.

While generalist teachers face challenges in teaching outside their area of expertise, the practice of out-of-field teaching can also present opportunities for professional growth and learning. Effective professional development and strong support from school leadership are critical in helping these teachers adapt and succeed in their roles, thereby ensuring quality education regardless of their primary specialisation (Crisan and Hobbs, 2019). Conversely, employing subject-qualified teachers remains a priority for many schools, as specialists bring focused expertise that enhances the learning experience in STEM subjects. To bridge these gaps, it is crucial to provide continuous professional development tailored to those teaching out-of-field, enabling both generalists and specialists to thrive in diverse STEM learning environments (Lyons et al., 2012; Du Plessis, 2020).

7.3 Framework for Sense Making

We employed the theory of practice architectures (Kemmis et al., 2014) to investigate the redesign of this STEM course. This approach helped us understand how the structural elements of the course shaped the learning practices of pre-service teachers and their educator. Studies in teacher education consistently demonstrate that career changers aspiring to be teachers, enter ITE programs with well-established beliefs about teaching and learning. These beliefs, highlighted by researchers such as Hogg et al (2023), often originate from their personal backgrounds and experiences. Simultaneously, initial teacher education programs are deeply entrenched in longstanding traditions shaped by specific narratives, practices, and hierarchical structures reflecting historical and cultural situatedness (Sjølie & Østern, 2021).

The theory of practice architectures as Kemmis et al. (2014) explains, is how people's actions are shaped by their social environment. This theory uses the idea of 'intersubjective space'—the areas where we interact with others—to show that our actions are shared and influenced by others. These actions happen in contexts where there are shared understandings, ways of doing things, and relationships of support and power.

Practice architectures include the structures within educational settings that shape and define actions. Kemmis et al. (2014) describe these structures as existing in three dimensions or arrangements: cultural-discursive (what is said), material-economic (what is done), and social-political (how people relate to each other). These structures influence how people talk, act, and interact within educational contexts.

Cultural-discursive arrangements involve the language and specialised discourse used in teaching and learning. For STEM education, this includes theoretical discussions about pedagogy, scientific concepts, and technical language used in classrooms and educational settings. Both pre-service teachers and initial teacher educators must navigate these arrangements to ensure a deep understanding of both general pedagogical practices and the specific STEM language, and concepts required.

Material-economic arrangements refer to the tangible resources and structures that support educational activities. This encompasses timetabling, the mandatory curriculum, availability of STEM equipment and facilities (such as teaching space and technology), and logistical aspects like lesson structure and planning. Initial teacher educators play an integral role in designing STEM courses that provide hands-on experiences and access to resources necessary for effective STEM teaching and learning. Pre-service teachers, in turn, engage with these resources to develop practical skills and apply theoretical knowledge in real-world contexts.

Social-political arrangements involve the roles, power dynamics, and organisational rules within educational institutions. This includes the distribution of authority, collaboration among educators and specialists, and the institutional norms that shape interactions and decision-making processes (Mahon et al., 2017). Both pre-service teachers and initial teacher educators need to navigate these arrangements to cultivate collaborative environments where learning can thrive.

These arrangements interact dynamically to create the landscape within which educational practices unfold. Employing practice architectures is a useful lens for analysing educational innovations, such as the development of new units or courses of work in Initial Teacher Education (ITE) programs. By examining the interplay of cultural-discursive, material-economic, and social-political arrangements, educators and policymakers can design and implement effective STEM education strategies that are responsive to the needs of both teachers and students (Mahon et al., 2017).

In the context of the multifaceted nature of teacher education programs, our inquiry will explore the research question—How do course design and interactions within the learning environment shape participant outcomes in a STEM teacher education course when analysed through the lens of practice architectures?

7.4 Research Approach

This study adopted a small-scale research paradigm (Knight, 2002) and was guided by an interpretivist stance (Elliott & Timulak, 2005) to explore the experiences of a cohort of postgraduate primary generalist pre-service teachers and their teacher educator during a series of immersive STEM experiences. In this compulsory postgraduate STEM education course, 28 primary pre-service teachers were enrolled and invited to participate in a concurrent research component while engaging in immersive STEM activities. These participants, all career changers, brought diverse professional backgrounds such as fashion design, law, marketing, and accounting to their postgraduate studies. Melanie, identified as Author 1, served as the teacher educator facilitating these experiences. With twenty-one years of experience in teacher education, Melanie brought a wealth of knowledge in science and STEM education.

As previously mentioned, the course under study was delivered intensively during the final semester of a two-year Master of Teaching Practice (Primary) program at

RMIT University, Melbourne, Australia. It included five one-day immersive workshops in September and October, with data collection occurring concurrently (see Table 7.1).

A descriptive case study methodology (Merriam, 1998) was used to guide the study, which consisted of three phases of data generation using a narrative inquiry approach (Merriam & Tisdell, 2016). Narrative inquiry is a foundational approach in qualitative education research (Parr et al., 2015) and was suitable for this study as it aimed to explore the lived experiences of the pre-service teachers and their educator in immersive STEM learning and teaching (Fig. 7.1).

Guided by our research question, the first phase involved gathering data from a variety of sources through an 'assemblage' approach (Strom, 2015) which combined data collected from pre-service teachers and the teacher educator. The pre-service teachers, insights were obtained via three sources:

1. A Likert survey assessing the pre-service teachers' confidence in teaching STEM;
2. A course evaluation survey consisting of Likert and open-ended questions;
3. A focus group discussion.

Table 7.1 Overview of pre-service teachers' second year immersive STEM course

Day	Session content
Day 1 University	Session 1: Introduction to the course and the immersive structure Session 2: Beginning the project Global Sustainability Goals 11 and 15, and Biodiversity Challenge Session 3: Critical reflection on activities and content—working on an integrated definition of STEM
Day 2 Tech school	Session 1: Explore innovative ideas for cities of the future futures through generative AI Session 2: Develop prototypes and gather feedback from peers Session 3: Critical reflection on activities and content—unpacking the design process and engineering concepts
Day 3 University	Session 1: Introduction to Esri Quick Capture app and developing biodiversity surveys Session 2: Conducting biodiversity surveys and analysing data Session 3: Critical reflection on activities and content—joining the dots between science and mathematics curricula
Day 4 University	Session 1: Revisiting our cities of the future prototypes and linking to the survey data collected Session 2: Group final planning for teaching episode to be conducted at the partner primary school and idea sharing Session 3: Feedback from teacher educator and peers on planned teaching episodes
Day 5 Primary School	Session 1: Teaching the 1st group of grades 5/6 at primary school—group reflection Session 2: Teaching the 2nd group of grades 5/6 at primary school—group reflection Session 3: Whole class critical reflection on teaching episodes and course wrap-up

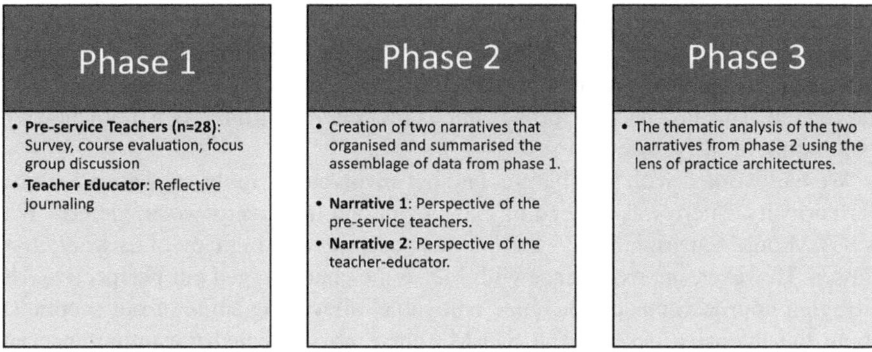

Fig. 7.1 Research phases

While reflective journaling was primarily used to capture Melanie's insights and her experiences working with various stakeholders. Upon completing the initial data collection, we analysed and shared the datasets among the three authors.

Inspired by Snow-Gerono et al. (2002), our second phase focused on crafting individual narrative vignettes to capture the perceptions of pre-service teachers and the teacher educator. We distilled each story into two narrative vignettes, presented earlier, identifying the most critical elements. This process involved collaborative coding by the three researchers and discussions to extract key elements for shaping the narratives.

In phase three, we analysed participant vignettes collaboratively. Using the theoretical framework of practice architectures (Kemmis et al., 2014), we examined the vignettes to identify themes related to cultural, economic, social, and political factors in teacher education. These themes were chosen to highlight influences on immersive STEM experiences. We used an inductive thematic approach (Bryman, 2016) to identify recurring patterns in the vignettes. To ensure reliability, the researchers discussed and agreed upon the themes collectively.

7.5 Stories of Practice

7.5.1 Pre-service Teachers' Vignette

When we first heard about the STEM course, our initial reactions were mixed. The news was unexpected, as STEM wasn't a natural fit for many of our career goals. For some of us, teaching STEM wasn't even a consideration in our vision of being primary teachers. In the schools where we conducted our professional practice, STEM was often outsourced to specialists or led by teachers with a genuine interest in the subject.

Additionally, the prospect of participating in another intensive course was not something we looked forward to. We had previously found the nature of intensive courses quite challenging. We felt overwhelmed by the sheer volume of material, rushing through topics that could have been explored in more depth. It often seemed like a race to complete assignments rather than an opportunity to engage deeply with the content. This fast-paced approach left us frustrated, yearning for a more thorough exploration of the subjects.

We had worked with Mel before, and her involvement in the new course piqued our curiosity. There was a blend of excitement and intrigue as we pondered, ‘Why is STEM our focus right now?’ STEM was not a specialism many of us would have chosen. However, our experience with Mel somewhat changed our perspective. Her engaging approach made the course enjoyable, alleviating some of our scepticism about the intensive design. The STEM course also revealed the importance and relevance of STEM, making us realise how valuable it can be to cultivate an interest in STEM in primary settings to prepare students to engage them in the world around them and for future evolving work opportunities.

Despite the intensive nature of the course and the fact that some of us had no initial interest in STEM, there were some positive elements that made the experience worthwhile. As pre-service teachers, we were particularly engaged by the brief dives into evolving technology like AI. We recognised the importance of understanding and teaching science in ways that connect with students, especially through real topics exploring environmental issues and disaster planning. Also, the hands-on practical lessons not only enhanced our understanding but inspired us to adopt similar methods in our future classrooms.

One of the most rewarding experiences was the final day at a local primary school. This practical application of STEM concepts has been a missing piece in our education program—not just in this course. We found our assignment wasn’t just about writing lesson plans for STEM, but we were enacting on them, seeing what worked in a real classroom setting. This opportunity to teach and then refine our plans based on real feedback from the primary students and Mel was invaluable. We all agreed that this kind of practical experience should have been integrated into the courses in our program much earlier.

Although appreciative of what the course offered and acknowledging that it differed from our initial expectations of an intensive, we desired greater variety. For example, exploring topics like planetary science, having more time to investigate assessment strategies, and focusing more on the E (Engineering) in STEM. Time constraints within the curriculum posed challenges in this regard. Nonetheless, our journey emphasised the significance of diverse and engaging teaching strategies in enhancing STEM education.

While the pre-service teachers who participated in this course all went on to be primary generalist teachers three did express a desire to become specialists in STEM/Science in the future and many more were willing to try science and STEM activities in their teaching.

7.5.2 Teacher Educator Vignette

In my capacity as the course coordinator, I was tasked with the responsibility of both designing and facilitating a curriculum tailored for the primary pre-service teachers within the second-year cohort. My initial engagement with this group involved conducting an intensive five-day introductory science course during their first year of study. Although the course structure was pre-established, I adapted to the situation and delivered the content as mandated. This first-year course was structured to compress the F-6 curriculum into a condensed time frame, entailing six hours of face-to-face instruction over 5 days.

Feedback from the participants indicated that while they loved the activities and material we covered, they were generally dissatisfied with the course's intensity. Taking this into consideration, and in anticipation of the subsequent intensive offering in the STEM course, I embarked on a strategic redesign of the Extending STEM course. The objective was to alleviate the burden of excessive content and instead, concentrate on cultivating the pedagogical competencies essential for effective STEM education. Such as designing and implementing authentic problem- and project-based hands-on activities, that facilitate learning through collaboration, critical thinking, and exploration.

To achieve this, I devised a novel approach for the five-day course structure that would engage pre-service teachers in a tripartite learning model: experiencing the roles of a learner, an educator, and a reflective practitioner. This model was aimed at helping the pre-service teachers to develop an understanding and practical application of STEM pedagogies. Despite the constraints of another intensive five-day schedule, I was determined to transform the course into a more engaging and immersive learning journey for the pre-service teachers. This reimagined format was designed to impart knowledge and inspire innovative teaching practices and reflective learning. I chose 2 of the UN's global sustainability goals on which to focus the course and came up with a 5-day program about biodiversity and cities of the future.

During the initial session of the course, attendance was notably below expectations, with a significant portion of the cohort absent due to the impending deadline of a capstone assignment task scheduled for the end of that week. Consequently, these pre-service teachers missed critical introductory elements, including an overview of the course content, operational logistics, and the foundational principles of the immersive structure designed to enhance their learning experience.

The following week's sessions, conducted at the Tech School separate from the university premises, were similarly impacted. Several pre-service teachers either arrived late or were absent for the second week in succession, which was particularly disconcerting given the week's focus on an intensive exploration of the design process and the application of technology in the development of innovative ideas and products. However, the third week marked a positive shift in attendance, aligning with the involvement of a specialist team from the geospatial software company Esri. Their contribution was instrumental in assisting the development of surveys intended for use in engaging children in the collection of citizen science data, thereby enriching the educational experience. Likewise, week 4 had good attendance as pre-service teachers planned and developed resources for their micro teaching session that would take place in a local primary school in Week 5.

The pre-service teacher feedback received regarding the micro-teaching session conducted in week 5 was overwhelmingly positive. Participants expressed a high level of satisfaction, appreciating the chance to engage in hands-on planning and execution of their ideas. This practical approach was favoured over the creation of a mere theoretical lesson sequence, typically assessed in academic settings.

One of the challenges was working with the other organisations and people from outside RMIT. Mainly the logistics of getting all the ducks in a row. Several times we had the school locked in for the microteaching and they changed the dates, this meant that all the other activities needed to change, but it worked out in the end. Some pre-service teachers not pulling their weight in the group's activities frustrated others. Also, the erratic attendance of some pre-service teachers was difficult.

My goals and future considerations involve deepening pre-service teachers' understanding of the course structure and clearly communicating its benefits. By doing so, I aim for pre-service teachers to see the value of attending all sessions and fully engaging with the course. This approach will enhance their grasp of the subject matter and ensure they benefit from the comprehensive learning experience designed to support their success. Ideally, this course will continue to be offered, but with potential changes in the curriculum due to reaccreditation, it may face a limited lifespan. If this occurs, I will need to adapt the content to fit a new structure while retaining its core objective and direction.

7.6 Unpacking the Stories Through 'Practice Architectures'

With the pre-service teachers' and teacher educator's narrative vignettes as our foundation, we now explore the themes that shaped both stories. The pre-service teachers and teacher educator vignettes both provide valuable insights into the experiences and challenges of implementing an intensive STEM course for primary pre-service teachers. By applying the theory of practice architectures, we can link these narratives and identify how cultural-discursive, material-economic, and social-political arrangements shape the practices in each context.

7.6.1 Cultural Discursive Arrangements

Cultural discursive arrangements shape how STEM education is perceived and valued within the educational context. The mixed initial reactions of the pe-service teachers to the STEM course highlights a broader cultural narrative that often positions STEM as peripheral to primary generalist teaching goals. This perception is not uncommon; studies indicate that primary educators often feel less confident and less prepared to teach STEM subjects compared to other areas of the curriculum (Adams et al., 2014; Lewis et al., 2021). This lack of confidence can stem from cultural norms and discourses that historically prioritise literacy and numeracy over STEM in primary education (Nadelson et al., 2013).

The pre-service teachers' experiences with STEM being outsourced to specialists reflect a discursive arrangement where STEM is viewed as a specialised domain. This separation can lead to a perception that STEM is beyond the capabilities of generalist primary teachers (Hobbs, 2012). However, the involvement of an engaging teacher like Mel helped to shift these perceptions, demonstrating that effective pedagogy can transform attitudes towards STEM. Research supports the notion that teacher enthusiasm and innovative instructional methods can significantly impact student engagement and perception of subjects (Glynn et al., 2007; Skinner et al, 2017).

The pre-service teachers' eventual recognition of the importance of STEM highlights a positive shift in cultural discourse. This aligns with broader educational goals to integrate STEM more holistically into the primary curriculum, promoting interdisciplinary teaching and relevance to real-world problems (Bybee, 2010; Nash et al, 2024). Cultivating interest in STEM at an early age is essential for developing future skills needed in an increasingly technological world (Honey et al., 2014). Thus, transforming cultural discourses around STEM in primary education is important for cultivating a generation of teachers and students who value and engage with STEM subjects.

7.6.2 Material Economic Arrangements

Material economic arrangements, such as the structure and delivery of the intensive STEM course, significantly impact educational experiences and outcomes. The pre-service teachers' frustration with the intensive format, characterised by condensed learning and high volume of material, highlights a common challenge in teacher education programs: balancing depth and breadth of content within limited timeframes (Darling-Hammond, 2006). Intensive courses can lead to superficial understanding and heightened stress, detracting from meaningful engagement with the material (Korthagen, 2017).

The reliance on external resources and specialists, such as the collaboration with GIS specialists in other faculties of the university and Esri, highlights the economic dimensions of STEM education. Access to technology and expertise can enrich the learning experience, but also poses logistical challenges. Research indicates that while partnerships with external organisations can provide valuable resources and expertise, they also require careful coordination and planning to align with educational objectives (Bell et al., 2014).

Hands-on practical experiences, like the final day at the primary school, are vital for bridging theory and practice. Such experiences enhance pre-service teachers' confidence and competence in teaching STEM (Palmer, 2006). However, the lack of earlier integration of these practical components reflects a material arrangement that often separates theoretical coursework from practical application that occurs within teacher education programs. Embedding more consistent and integrated practical experiences throughout teacher education programs can mitigate this gap, promoting a more cohesive and effective learning environment (Zeichner, 2010).

7.6.3 Social Political Arrangements

Social political arrangements influence how STEM education is structured and implemented within teacher education programs. The teacher educator's strategic redesign of the course to include a tripartite learning model—learner, educator, and reflective practitioner—reflects an effort to align with contemporary educational frameworks that emphasise holistic and reflective teaching practices. This approach is supported by literature advocating for reflective practice to enhance teaching effectiveness and professional growth (Fullard & Watts, 2022; Thompson & Pascal, 2012).

The inclusion of global sustainability goals and collaboration with external organisations highlights the social political drive towards interdisciplinary and socially relevant education. Integrating topics like biodiversity and cities of the future not only makes STEM more engaging but also aligns with educational policies promoting sustainability and real-world problem solving (Anderson, 2012; UNGA, 2015). However, logistical challenges, such as inconsistent attendance and coordination with external partners, illustrate the complexities of implementing such innovative approaches.

Feedback indicating pre-service teachers' appreciation for practical, hands-on experiences over theoretical assignments suggests a need for teacher education programs to put more emphasis on experiential learning. This aligns with constructivist theories that advocate for learning through doing (Kolb, 1984; Morris, 2020; Seaman et al., 2017). Ensuring that pre-service teachers understand the rationale behind course structures and the importance of consistent participation can only enhance their engagement and learning outcomes. Addressing these social political arrangements can lead to more effective and sustainable STEM education practices within teacher education programs (Cochran-Smith & Zeichner, 2005).

7.7 Conclusion

In examining how course design and interactions within the learning environment shape participant outcomes in a STEM teacher education unit through the lens of practice architectures, we can see that cultural-discursive, material-economic, and social-political arrangements play pivotal roles.

Culturally, pre-service teachers initially viewed STEM as peripheral to their primary teaching roles, shaped by broader educational narratives that prioritise literacy and numeracy. However, through engaging pedagogy and positive role models, there was a discursive shift in their attitudes, reflecting the potential to transform STEM's position in primary education. This shift suggests that pre-service teachers may be more inclined to integrate STEM into their future classrooms, recognising its importance and becoming more confident in their ability to teach STEM concepts.

Materially, the intensive course structure presented challenges in terms of workload and depth of learning, highlighting the importance of balancing content delivery with meaningful engagement. Partnerships with external resources, while beneficial, also required careful logistical planning to integrate effectively with the course objectives. These partnerships were key in demonstrating how a primary generalist might draw on community support in their own teaching situation, offering practical examples of how external expertise and local resources can be leveraged to enrich the learning experience and provide real-world context. This collaboration not only enhanced the relevance of the course content but also modelled effective strategies for creating resourceful, community-oriented classroom environments.

Socially and politically, the course's design, particularly the emphasis on reflective practice and interdisciplinary, real-world problems, aligns with modern educational frameworks that promote sustainability and practical, hands-on learning. Yet, the logistical issues faced demonstrate the need for clearer communication and more consistent participation in such innovative programs.

Overall, the analysis suggests that course design and interactions within the learning environment can impact participant outcomes in STEM teacher education. By addressing cultural, material, and social factors, a course can challenge existing narratives, offer practical experiences, and cultivate a positive attitude towards STEM teaching. Carefully considering these factors could help in equipping pre-service generalist teachers with the knowledge, skills, and attitudes needed to integrate STEM into their future classrooms.

References

Adams, A. E., Miller, B. G., Saul, M., & Pegg, J. (2014). Supporting elementary pre-service teachers to teach STEM through place-based teaching and learning experiences. *The Electronic Journal for Research in Science & Mathematics Education, 18*(5).

Anderson, A. (2012). Climate change education for mitigation and adaptation. *Journal of Education for Sustainable Development, 6*(2), 191–206. https://doi.org/10.1177/0973408212475199

Australian Institute for Teaching and School Leadership. (2021). *Teacher supply and demand report 2021*. AITSL.

Bell, P., Lewenstein, B., Shouse, A. W., & Feder, M. A. (Eds.). (2014). *Learning science in informal environments: People, places, and pursuits*. National Academies Press.

Bryman, A. (2016). *Social research methods*. Oxford University Press; Center for Engaged Learning. (2024). *Immersive learning*. Elon University. Retrieved from https://www.Centerforengagedlearning.org/resources/immersive-learning/

Bybee, R. W. (2010). Advancing STEM education: A 2020 vision. *Technology and Engineering Teacher, 70*(1), 30–35.

Cochran-Smith, M., & Zeichner, K. M. (Eds.). (2005). *Studying teacher education: The report of the AERA panel on research and teacher education*. Lawrence Erlbaum Associates.

Collins, A., Hay, I., & Heiner, I. (2013). Start with the end in mind: Experiences of accelerated course completion by pre-service teachers and educators. *Australian Journal of Teacher Education* 38(10), 1–20. https://search.informit.org/doi/https://doi.org/10.3316/ielapa.680020102673874

Crisan, C., & Hobbs, L. (2019). Subject-specific demands of teaching: Implications for out-of-field teachers. In L. Hobbs & G. Törner (Eds.), *Examining the phenomenon of "teaching out-of-field"*. Springer. https://doi.org/10.1007/978-981-13-3366-8_6

Darling-Hammond, L. (2006). Constructing 21st-century teacher education. *Journal of Teacher Education, 57*(3), 300–314.

Du Plessis, A. E. (2020). The lived Experience of out-of-field STEM teachers: A Quandary for strategising quality teaching in STEM? *Research in Science Education, 50*, 1465–1499. https://doi.org/10.1007/s11165-018-9740-9

Elliott, R., & Timulak, L. (2005). Descriptive and interpretive approaches to qualitative research. In J. Miles & P. Gilbert (Eds.), *A handbook of research methods for clinical and health psychology* (pp. 147–159). Oxford University Press

Fullard, M., & Watts, P. M. (2022). Practical wisdom in teaching through the promotion of critical reflection during ITE. In *Encyclopedia of teacher education* (pp. 1351–1355). Springer Nature Singapore.

Gilstrap, D. L. (2013). Why do we teach? Adult learning theory in professional standards as a basis for curriculum development. *College & Research Libraries, 74*(5), 501–518. https://doi.org/10.5860/crl12-334

Glynn, S. M., Brickman, P., Armstrong, N., & Taasoobshirazi, G. (2007). Science motivation questionnaire: Construct validation with nonscience majors. *Journal of Research in Science Teaching, 44*(6), 527–536. https://doi.org/10.1002/tea.20267

Hobbs, L. (2012). Examining the aesthetic dimensions of teaching: Relationships between teacher knowledge, identity and passion. *Teaching and Teacher Education, 28*(5), 718–727. https://doi.org/10.1016/j.tate.2012.01.010

Hogg, L., Elvira, Q., & Yates, A. (2023). What can teacher educators learn from career-change teachers' perceptions and experiences: A systematic literature review. *Teaching and Teacher Education, 132*, Article 104208. https://doi.org/10.1016/j.tate.2023.104208

Honey, M., Pearson, G., & Schweingruber, H. (Eds.). (2014). *STEM integration in K-12 education: Status, prospects, and an agenda for research*. National Academies Press.

Hurley, M., Butler, D., & McLoughlin, E. (2024). STEM teacher professional learning through immersive STEM learning placements in industry: A systematic literature review. *Journal for STEM Education Research, 7*(1), 122–152. https://doi.org/10.1007/s41979-023-00089-7

Kemmis, S., Wilkinson, J., Edwards-Groves, C., Hardy, I., Grootenboer, P., & Bristol, L. (2014). *Changing practices, changing education*. Springer.

Kirkiç, K. A., & Arikan, E. E. (2023). Primary school teachers' attitudes and views toward STEM education. *Science Education International, 34*(2), 132–141. https://doi.org/10.33828/sei.v34.i2.7

Knight, P. T. (2002). *Small-scale research: Pragmatic inquiry in social science and the caring professions*. Sage

Kolb, D. A. (1984). *Experiential learning: Experience as the source of learning and development*. Prentice-Hall.

Korthagen, F. A. J. (2017). Inconvenient truths about teacher learning: Towards professional development 3.0. *Teachers and Teaching, 23*(4), 387–405. https://doi.org/10.1080/13540602.2016.1211523

Lange, A. A., Nayfeld, I., Mano, H., & Jung, K. (2022). Experimental effects of a preschool STEM professional learning model on educators' attitudes, beliefs, confidence, and knowledge.

Journal of Early Childhood Teacher Education, 43(4), 509–539. https://doi.org/10.1080/10901027.2021.1911891

Lewis, F., Edmonds, J., & Fogg-Rogers, L. (2021). Engineering science education: The impact of a paired peer approach on subject knowledge confidence and self-efficacy levels of student teachers. *International Journal of Science Education, 43*(5), 793–822. https://doi.org/10.1080/09500693.2021.1887544

Lyons, T., Quinn, F., Rizk, N., Anderson, N., Hubber, P., Kenny, J., Sparrow, L., & Wilson, S. (2012). *Starting out in STEM: A study of young men and women in first year science, technology, engineering and mathematics courses.* SiMERR National Research Centre, University of New England.

Mahon, K., Francisco, S., & Kemmis, S. (2017). *Exploring education and professional practice: Through the lens of practice architectures* (1st ed.). Springer. https://doi.org/10.1007/978-981-10-2219-7

Merriam, S. B. (1998). Qualitative research and case study applications in education. Jossey-Bass Publishers.

Merriam, S. B., & Tisdell, E. J. (2016). *Qualitative research: A guide to design and implementation.* Wiley.

Morris, T. H. (2020). Experiential learning: A systematic review and revision of Kolb's model. *Interactive Learning Environments, 28*(8). https://doi.org/10.1080/10494820.2019.1570279

McDonald, C. V. (2016). STEM Education: A review of the contribution of the disciplines of science, technology, engineering and mathematics. *Science Education International, 27*(4), 530–569.

Nadelson, L. S., Callahan, J., Pyke, P., Hay, A., Dance, M., & Pfiester, J. (2013). Teacher STEM perception and preparation: Inquiry-based STEM professional development for elementary teachers. *Journal of Educational Research, 106*(2), 157–168. https://doi.org/10.1080/00220671.2012.667014

Nash, M., Bradbury, O., & Fitzgerald, A. (2024). Understanding the perspectives of a teacher educator and pre-service teachers toward an immersive STEM experience. *Future in Educational Research*, 1–19. https://doi.org/10.1002/fer3.53

Palmer, D. (2006). Sources of self-efficacy in a science methods course for primary teacher education students. *Research in Science Education, 36*(4), 337–353.

Parr, G., Bulfin, S., Castaldi, R., Griffiths, E., & Manuel, C. (2015). On not becoming 'a mere empirical existence': Exploring 'who' and 'what' narratives in pre-service English teachers' writing. *Cambridge Journal of Education, 45*(2), 133–148. https://doi.org/10.1080/0305764x.2014.930416

Pimthong, P., & Williams, P. J. (2021). Methods course for primary level STEM pre-service teachers: Constructing integrated STEM teaching. *EURASIA Journal of Mathematics, Science and Technology Education, 17*(8). https://doi.org/10.29333/ejmste/11113

Razi, A., & Zhou, G. (2022). STEM, iSTEM, and STEAM: What is next? *International Journal of Technology in Education, 5*(1), 1–29. https://doi.org/10.46328/ijte.119

Rinke, C. R., Gladstone-Brown, W., Kinlaw, C. R., & Cappiello, J. (2016). Characterizing STEM teacher education: Affordances and constraints of explicit STEM preparation for elementary teachers. *School Science and Mathematics, 116*(6), 300–309. https://doi.org/10.1111/ssm.12185

Sahito, Z., & Wassan, S. H. (2024). Literature review on STEM education and its awareness among teachers: An exploration of issues and problems with their solutions. *SAGE Open, 14*(1). https://doi.org/10.1177/21582440241236242

Seaman, J., Brown, M., & Quay, J. (2017). The evolution of experiential learning theory: Tracing lines of research in the JEE. *Journal of Experiential Education, 40*(4). https://doi.org/10.1177/1053825916689268

Sjølie, E., & Østern, A. L. (2021). Student teachers' criticism of teacher education–through the lens of practice architectures. *Pedagogy, Culture & Society, 29*(2), 263–280. https://doi.org/10.1080/14681366.2020.1729842

Skinner, E., Saxton, E., Currie, C., & Shusterman, G. (2017). A motivational account of the undergraduate experience in science: Brief measures of students' self-system appraisals, engagement in coursework, and identity as a scientist. *International Journal of Science Education, 39*(17), 2433–2459. https://doi.org/10.1080/09500693.2017.1387946

Snow-Gerono, J. L., Yendol-Silva, D., & Nolan, J. F. (2002). Reconceptualizing curriculum for the PDS: University faculty negotiate tensions in collaborative design of methods courses. *Action in Teacher Education, 24*(3), 63–72. https://doi.org/ https://doi.org/10.1080/01626620.2002.10734432

Strom, K. (2015). Teaching as assemblage: Negotiating learning and practice in the first year of teaching. *Journal of Teacher Education, 66*(4), 321–333. https://doi.org/10.1177/0022487115589990

Thompson, N., & Pascal, J. (2012). Developing critically reflective practice. *Reflective Practice, 13*(2), 311–325. https://doi.org/10.1080/14623943.2012.657795

United Nations General Assembly [UNGA]. (2015). Transforming our world: The 2030 agenda for sustainable development. Wang, C., Cho, H. J., Wiles, B., Moss, J. D., Bonem, E. M., Li, Q., Lu, Y., & Levesque-Bristol, C, 2022 United Nations General Assembly [UNGA]. (2015). Transforming our world: The 2030 agenda for sustainable development; Wang, C., Cho, H. J., Wiles, B., Moss, J. D., Bonem, E. M., Li, Q., Lu, Y., & Levesque-Bristol, C. (2022). *Competence and autonomous motivation as motivational predictors of college students' mathematics achievement: From the perspective of self-determination theory. IJ STEM Ed, 9*(1), 41. https://doi.org/10.1186/s40594-022-00359-7

Weldon, P. (2016). *Out-of-field teaching in Australian secondary schools.* (Policy Insights; n.6). Australian Council for Educational Research (ACER).

White, B., MacGregor, D. D., & Panizzon, D. D. (2018). *Developing pre-service teachers' understandings of STEM through an industry experience.* In *Post Conference Proceedings for the 5th International STEM in Education Conference* (pp. 437–443). Retrieved 29 July, 2024, from https://researchmgt.monash.edu/ws/portalfiles/portal/263150128/262939822_oa.pdf

Zeichner, K. (2010). Rethinking the connections between campus courses and field experiences in college- and university-based teacher education. *Journal of Teacher Education, 61*(1–2), 89–99.

Chapter 8
Primary STEM Education in the Netherlands

Ronald Keijzer, Anika Embrechts, Anna Hotze, Vincent Jonker, Margien Bakker, Lonneke Boels, and Symen van der Zee

Abstract This chapter elaborates on three challenges related to the specific character of Science, Technology, Engineering and Mathematics (STEM) education in primary education. It describes the SPRONG STEM project results. The SPRONG STEM project aimed at strengthening STEM education infrastructure in The Netherlands by developing professional learning networks, wherein teachers and researchers participate. Before focusing on these challenges, we sketch out an overview indicating the importance of STEM education. From this partly historical analysis we focus on the following challenges: (1) assessing learning outcomes in STEM education (Sect. 8.2), (2) the relation between mathematics and other STEM domains (Sect. 8.3), and (3) data literacy and maker education in STEM education (Sect. 8.4). With respect to these challenges we describe both theoretical as well as practical questions and demands that emerged from professional learning networks, including the critical

R. Keijzer (✉) · M. Bakker
IPABO University of Applied Sciences, Amsterdam, The Netherlands
e-mail: r.keijzer@ipabo.nl

M. Bakker
e-mail: m.bakker@ipabo.nl

A. Embrechts
Regional Community College of Twente, Hengelo, The Netherlands
e-mail: aembrechts@rocvantwente.nl

A. Hotze
Delft University of Technology, Delft, The Netherlands
e-mail: A.Hotze@tudelft.nl

V. Jonker
Utrecht University, Utrecht, The Netherlands
e-mail: v.jonker@uu.nl

L. Boels
Hogeschool Utrecht, University of Applied Sciences, Utrecht, The Netherlands
e-mail: Lonneke.Boels@hu.nl

S. van der Zee
Saxion University of Applied Science, Enschede, The Netherlands
e-mail: s.vanderzee@saxion.nl

S. Earle et al. (eds.), *Primary Science Learning for Children, Teachers, and Communities*, https://doi.org/10.1007/978-981-95-0561-6_8

role of teachers' content knowledge. Classroom examples further illustrate these challenges. The chapter concludes with recommendations for further research and curriculum development in STEM education.

Keywords STEM education · Assessment · Professional development · Professional learning networks

8.1 STEM Education in the Netherlands

8.1.1 Importance of STEM Education

Understanding complex present-day problems like climate change or migration touches upon all four domains in Science, Technology, Engineering and Mathematics (STEM). For example, climate change is explained using sciences such as chemistry and physics, while mathematics helps to model the situation. Technology and engineering are required to deter climate change, for example, by designing and manufacturing cars and airplanes that run on hydrogen. STEM education is considered essential to solving these challenges and therefore important to Dutch society (Bom et al., 2019; Gresnigt, 2018; Rohaan, 2009). The increasing shortage of workers in the STEM fields has a negative impact on economies. In addition, technologization is rapidly changing the labor market. More and more jobs require scientific, mathematical and technological competencies (Gal, 2024a; Hoogland, 2023; Kirschner & Stoyanov, 2020). Moreover, rapid scientific and technological advancements raise social and ethical issues. Understanding mathematics, science and technological developments, their possibilities, limits, and the ethical and democratic dilemmas that they bring forth is crucial for every citizen in today's society (Gravemeijer et al., 2017; Guérin, 2018; Laugksch, 1999). Primary education should therefore acquaint children from early on with STEM, develop their STEM content knowledge, skills, and attitudes, and stimulate them to continue learning about STEM (Post, 2019).

8.1.2 Historical Perspective

Over the past two decades, there have been repeated calls in the Netherlands to increase attention to STEM in primary education. In Dutch primary education the STEM areas are often referred to as Science and Technology (S&T) education. Mathematics has its own curriculum and learning goals. Although repeated calls have been made to connect Mathematics and S&T, these subjects are usually taught separately in primary classrooms. Until recently, the S&T curriculum mainly emphasized life science and physical science. Technology constituted only a small part of the curriculum. However, due to a growing focus on technology in society, technology and engineering (T&E) is more explicitly incorporated into science, resulting in a more balanced S&T(E) curriculum (Rohaan, 2009). This also means increased attention to engineering, as engineering knowledge and skills are an integral part of technology education (Post, 2019).

Initially, the S&T(E) curriculum focused on developing content knowledge and skills. A vision report published in 2005 introduced a new dimension: stimulating S&T(E) attitudes by nurturing children's inquiring minds (Post, 2019). Children with a positive attitude towards S&T(E) are more inclined to pursue a career in STEM fields (Bom et al., 2019). Therefore, it is important that teachers stimulate the development of a positive attitude (Van Aalderen-Smeets & Walma van der Molen, 2013). Stimulating this attitude towards S&T, teachers are encouraged to structure their science lessons according to the inquiry learning cycle and their technology lessons according to the design learning cycle, stimulating students to think and work as scientists and engineers (Bom et al., 2019; Rohaan, 2009).

Although many (prospective) teachers enrolled in S&T(E) courses, the number of schools implementing design and inquiry learning remained limited. According to studies conducted at that time, attention to S&T(E) education actually decreased (Kneepkens et al., 2011). An Advisory Committee on Science and Technology Education was established to advise on integrating S&T(E) into the primary school curriculum. By spring 2013, the committee issued its recommendations. One suggestion was to have the National Institute for Curriculum Development in the Netherlands, SLO, develop the S&T(E) curriculum and exemplary learning materials for primary education. Additionally, the committee proposed integrating S&T(E) with other subjects, particularly mathematics and language (Post, 2019).

A national S&T(E) curriculum was published in 2018 and operates on the premise that S&T(E) is a lens through which to view and approach the world. By posing questions, solving problems, and using imagination, children learn about the world, develop their research and engineering skills, and develop their scientific attitude. Despite offering practical insights, the implementation of the S&T(E) curriculum has fallen short in practice (Post, 2019). A recent study by Djoyoadhiningrat-Hol and Klein Tank (2023) indicates that less time is being allocated to S&T(E). While most schools aspire to implement the S&T(E) curriculum and work on it structurally, there is no explanation for this finding.

8.1.3 Connecting S&T(E) and Mathematics

Although mathematics has its own developmental history, curriculum and attainment goals in Dutch education (Van den Heuvel-Panhuizen, 2020), schools as well as experts in the field of mathematics teaching express a desire to connect S&T(E) with mathematics (Van der Aalsvoort et al., 2020). Some of the major reasons for this are the increasing role of mathematics in other fields and the approach to primary mathematics education in the Netherlands called Realistic Mathematics Education (RME). RME takes meaningful mathematical situations as a starting point and supports students in a process of mathematizing, allowing children to start learning from their intuitive mathematical notions and informal procedures. Under the careful guidance of a teacher, children rediscover and reconstruct mathematics (Oonk et al., 2020). This socio-constructivist approach to mathematics learning lends itself to connecting with other subjects of the curriculum and curriculum integration. Thus, mathematics need not be taught only as a separate subject.

Some of research and engineering skills, like modelling and representing are part of the S&T(E) curriculum and also match aspects of the mathematics curriculum. This however is not enough for teachers to integrate mathematics with other ST(E)M subjects. Experts agree that primary school teachers lack subject matter knowledge, pedagogical content knowledge, and self-efficacy to teach S&T(E) and to connect it in meaningful ways to mathematics (Bakker et al., 2023; Djoyoadhiningrat-Hol & Klein Tank, 2023; Hotze & Keijzer, 2017;). This aligns with various national and international research studies which indicate that proper curriculum integration requires far reaching expertise from teachers (Gresnigt, 2018; Van der Aalsvoort et al., 2020) and studies which indicate a general lack of knowledge, skills, and confidence among primary school teachers of S&T(E) (Bom et al., 2019; Davis et al., 2006; Post, 2019).

8.1.4 Professional Learning Networks

As science education is often seen as relatively new and innovative in Dutch schools, experienced teachers tend to rely on professional learning networks. Often these networks developed from a professionalization trajectory in science education, where teachers generally learned about specific characteristics of science education. Participants in the trajectory felt the need to implement a science curriculum in their schools and wanted to learn from each other. Examples of practice in this chapter come from such professional learning networks (PLNs). They consist of teachers, expert teachers in one of the STEM domains, researchers, and teacher educators. These networks, for example, support teachers to experience the complexity of inquiry- and design-based learning, and identify the gaps between such learning and the textbooks they use. A majority of school use textbooks for STEM, however these textbooks are focused on geography, history and science only (Djoyoadhiningrat-Hol & Klein Tank, 2023). Participants learn within their PLN, but also share findings with colleagues who do not participate in the PLN. This facilitates knowledge sharing between STEM experts and teachers, within and beyond the PLNs. For example, classroom observation instruments and tools for inquiry and design-based teaching skills are shared (Van Graft & Klein Tank, 2018).

8.1.5 Recurring and New Challenges

The brief history of STEM in Dutch primary education reveals several recurring challenges. The first is ensuring that teachers possess sufficient understanding of S&T(E) to deliver high-quality education. As mentioned, many primary school teachers lack the knowledge, skills and attitudes for teaching STEM. The second issue concerns connecting S&T(E) and mathematics. Mathematics is often taught separately from S&T(E). In fact, experts indicate that the increased attention on mathematics in recent years has replaced attention to S&T(E). Given the possibilities to connect and even integrate mathematics with STEM, this is a highly undesirable situation. The question is, therefore, how mathematics and S&T(E) can be connected in educational practice in meaningful and feasible ways. These issues remain unresolved and are therefore also described below as challenges.

Another demanding current issue concerns the new learning goals. In 2022, the Ministry of Education, Culture and Science commissioned the Institute for Curriculum Development (SLO) to update all attainment goals for primary and lower secondary education. The objective was to align the goals with societal developments and to establish a more coherent curriculum. To keep the curriculum up-to-date, new attainment goals were formulated for two new domains: digital literacy and citizenship education. The goals in both these domains exhibit significant overlaps with the updated attainment goals for mathematics and S&T(E). The plan is for all learning goals to be developed by 2024, followed by a 12–18 month period of implementing and testing their usability in practice. The question now confronting STEM educators is: How can these goals be meaningfully interwoven into the S&T(E) and mathematics curriculum?

Following the partly historical analysis in Sect. 8.1, in the next sections, we focus on four challenges: (1) assessing learning outcomes, (2) the relation between mathematics and the other STEM domains, and (3) Maker literacy and data literacy in STEM education.

In the SPRONG-STEM project, funded by the National Coordinating Institute for Education Research (NRO), we addressed three of these challenges by setting up local and national professional learning networks (PLNs). Below we describe the theoretical considerations as well as practical implications and issues in classrooms that emerged from these professional learning networks.

8.2 Assessing Learning Outcomes

8.2.1 The Challenge of Assessing S&T(E)

In S&T(E) education students develop new knowledge and skills (inquiry, design and engineering), and work on attitudinal aspects simultaneously. With increased implementation of inquiry and design in S&T(E) education, teachers feel the need to assess these aspects of learning as well. In a typical S&T activity, students are stimulated to use their curiosity and think critically. They gain new insights through research or working on a design project, learn to pose questions, make a research plan, collect evidence, compare results, and draw conclusions. The role of these process skills in developing a thorough understanding of phenomena is crucial (Harlen, 1999). It is therefore essential to assess these science and engineering process skills as well as the acquired content knowledge. In the Netherlands, schools and teachers are obliged to work on national key learning objectives. For example, one key learning objective for S&T is: 'students learn to do research on materials and physical phenomena, such as light, sound, electricity, force, magnetism and temperature' (SLO, 2006). The national institute for curriculum development in the Netherlands further elaborates these key learning objectives for different grade levels in primary education with respect to both knowledge and skills. Teachers often assess students' learning outcomes with respect to content knowledge through a test or quiz at the end of each theme in most Dutch textbooks. However, assessments for science research and design skills are generally not included in these textbooks. A possible reason may be the complexity of skills involved in learning design and inquiry (Table 8.1). Skills involved in design are broad (Crismond & Adams, 2012). These skills are therefore difficult to assess. The same is true of inquiry skills. Pedastre et al. (2015) developed an enhanced framework showing all aspects of inquiry-based teaching and learning.

The complexity of inquiry- and design-based learning means that teachers generally experience difficulty assessing learning outcomes with respect to these skills. Moreover, these skills cannot be isolated from the content or subject matter of science and engineering. In the following section, we present examples of primary school teachers working in PLNs to improve science education and assessment of learning outcomes. In one PLN of two teachers, a school policy adviser and a researcher, an assessment tool for inquiry and design-based learning has been developed. A question discussed in the PLN was whether to assess children during the process of inquiry or design, or at the end, based on their final thesis or presentation. Another question was how to incorporate all different aspects of inquiry or design but not make it too complicated to fill in the assessment tool. Therefore, in a second PLN

Table 8.1 Skills involved in inquiry and design in primary education

Inquiry skills	Design skills
Identifying investigable questions	Problem exploration
Designing investigations	Generating ideas, formulate design requirements
Obtaining evidence	Create a working prototype
Interpreting evidence with respect to research question	Testing and optimization
Communication the results of the investigation process	Presenting

a tool was developed to analyze the quality of the S&T activities carried out by teachers, focusing on various aspects of S&T.

8.2.2 Classroom Examples of Assessing Learning Outcomes

In S&T(E) education teachers often use textbooks. They discuss the use of textbooks in the PLNs. Teachers are encouraged to adapt activities in the textbooks and thereby explore new ways to make learning outcomes visible. They also express the need for tools to help them analyze the quality of S&T activities and improve them. Table 8.2 shows an example observation list for analyzing the quality of S&T-activities carried out by teachers in early childhood education.

For example, a S&T(E) textbook used in a school in the northern part of the Netherlands includes tests at the end of each theme to assess students' S&T(E) content knowledge. However, this textbook does not provide tools for assessing students' S&T(E) skills. Therefore, a PLN consisting of teachers and researchers collaborated to develop a tool to assess learning outcomes. They based their tool (Table 8.2) on the learning trajectories of inquiry and design in primary education developed by the National Institute for Curriculum Development in the Netherlands (SLO) (Van Graft & Klein Tank, 2018). Teachers in the PLN indicated that the tool should enable both teachers and students to assess learning outcomes. They operationalized learning outcomes of using steps of inquiry- and design-based learning into more practical and recognizable goals for students. In the process teachers agreed that the tool should be a score sheet, which could be easily used by both teachers and students. This resulted in a score sheet based on the sheets developed by Van Keulen and slot (2013) for research but adapted for use in teaching. Categories in the sheet included: recognizing the problem, developing a solution, testing and improving the design, and presenting the design.

In the other PLN, five teachers, two national curriculum advisors, one school policy advisor and one researcher worked on S&T content knowledge and skills in regard to the quality of S&T activities. A specific area of interest was for instance S&T in early childhood education. Questions discussed in this PLN were how to support teachers to expand their curriculum awareness and reveal embedded STEM

Table 8.2 Tool developed in a PLN for analyzing S&T activities in early childhood education

S&T(E) observation instrument
1. Is the activity embedded in a broader context?
2. Is student's curiosity and inquiry supported?
3. Is content from other domains, like geography or history, involved in the activity?
4. Are students developing S&T reasoning skills?
5. Are generic skills, like language or mathematics, involved?
6. Extra: do students use real materials?

skills in current classroom activities. In this PLN a tool was developed to analyze the quality of the S&T activities carried out by teachers.

8.2.3 *Reflection*

Many teachers still experience challenges teaching S&T(E). However, teachers participating in PLNs and S&T(E) expert teacher groups did find a way to narrow this knowledge gap through joint practice-based research projects. A common theme in these PLNs is improving science education by better observing and assessing students' knowledge and skills before, during and after S&T activities. This resulted in several observation sheets and scoring methods being developed by teachers.

8.3 The Challenge of Connecting S&T(E) and M

8.3.1 *Setting the Scene*

The four STEM perspectives are simultaneously needed to understand many current events. However, primary school teaching practice integrating the four disciplines is rare. In our experience as experts in the fields of mathematics and S&T(E) we notice that in most primary school curricula integrated STEM-activities have either a S&T(E) focus or a mathematics focus. In mathematics education, for example, there is a focus on mathematical procedures, and the scientific context provided is often irrelevant. By contrast, in S&T(E) teaching the embedded mathematics is usually just a tool providing (instrumental) procedures to work with numbers. In solving this problem of limited focus in integrating STEM domains, Lonning and DeFranco (2010) describe a continuum of 'independent mathematics—mathematics focus—balanced mathematics and science—science focus—independent science'. However, this continuum does not support simultaneous learning in all four domains. One reason that integration of these domains is problematic is that the domains are structured differently (Hotze & Keijzer, 2017). Key to mathematics are activities like symbolizing, modelling, formalizing, problem solving, pattern recognition, and generalizing. On the other hand, science focuses on explaining phenomenon in nature, using instruments and technology to do so. As a consequence, mathematics education sets the scene for mathematizing the world, while S&T education's focus is on research activities, designing, and explaining.

Several educational designers provide examples where integration of the STEM domains was successful in the sense that there was equal focus on two or more STEM-domains. These researchers, for example, showed that modelling with data, problem solving and data analysis were generative contexts to integrate science and mathematics in primary education (English, 2015; Mulligan & English, 2014).

Problem solving, working with data, and modelling also appear in most frameworks for so-called twenty-first century skills (Voogt & Pareja Roblin, 2012). Moreover, dealing with data and modelling also typifies the mathematics needed for the future (Gal, 2024b; Gravemeijer et al., 2017). These focal points are also included in the draft of the Dutch national standards for mathematics in primary education (Prenger et al., 2023).

The experiences of the PLNs however show that integrating the STEM domains in teaching places high demands on teachers' domain specific content knowledge and pedagogical content knowledge (Maass et al., 2019).

8.3.2 Example: 'Growing Grain for Bread'

In a PLN in the Amsterdam region, two teachers and two researchers developed an activity where science and mathematics are integrated for grade 4 (an age-group of approximately 10 years). The two teachers were specialists with a background in engineering and biology. While exploring possible activities in the PLN, teachers and educators in the PLN, first focused on mathematical thinking and mathematizing. They took the perspective of a specific domain and then changed this perspective—for example from mathematics to science—as a way to integrate both domains in the activity. An example for this changing in perspective is in exploring a bicycle from a technological point of view first. This might focus on the chain and how this enables actual biking. Next a mathematical perspective can lead to schematizing the ratio between blades the chain runs by. While sharing and discussing this idea of changing perspective with the two teachers in the PLN, educators designed a learning environment for teachers, where they learned to change perspectives from one domain to the other, and an activity for primary school students where both domains were integrated.

The activity for the students focused on domain specific thinking and changing perspective from one domain to the other. After discussing this idea, one of the teachers in the PLN developed the activity 'The journey from grain to bread' (Bakker et al., 2023). Bread forms an important part of the daily diet for many children in the Netherlands. The development from grain to bread provides sufficient opportunities for students to explore the situation, like from sowing to milling and baking bread. Moreover, this trajectory from grain to bread requires both mathematical thinking, like estimating the number of grains required for one loaf, and insights from biology, like how the grain grows.

When the teacher introduces the context of grain and bread, she starts by posing an open question: How do you get from grain to bread? In their responses students show what they know about sowing, harvesting and milling. After exploring the situation, the teacher introduces a more focused problem: 'Suppose we use the school garden to grow wheat for bread, how many loaves of bread could we make?' Students are invited to share their estimates: fifteen, three, less than one. The teacher replies: 'And how can we figure this out?'.

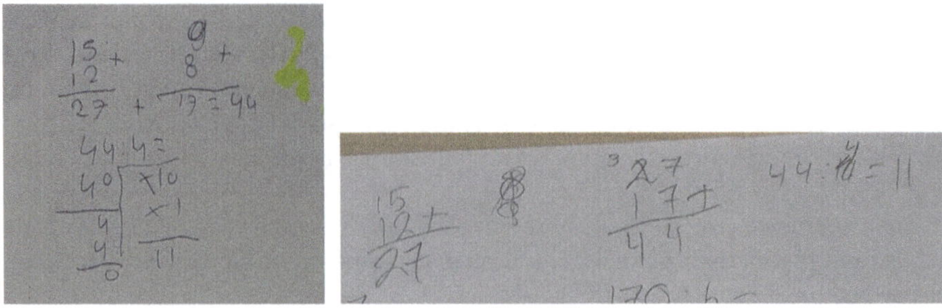

Fig. 8.1 Student work of stalks per plant (Photos by authors)

The children come up with different approaches to solve the problem, starting with how much grain a loaf of bread requires or starting with how much grain it is possible to grow in the school garden. As most students favor the latter option, the teacher presents a map of the school garden and together with the students explores the garden's area. When this is settled students formulate additional subproblems to solve the original problem. In the process they change perspective from biology, namely how grain grows, to mathematics, the number of stalks per plant (Fig. 8.1).

The students investigate the multiplicative structure of the problem and calculate, moving from the number of stalks per plant to the number of grains per stalk. They do this by counting and schematizing the result to help add the numbers. Multiplication emerges here as repeated addition.

Actual grain available during the activity allows students to investigate the grain as a substance, by feeling, tasting and grinding it. They also share their observations about the smell, the taste and strength of the grain. Eventually, the ground grain, in the form of flour, is weighed. When the students finally combine their mathematical findings, they conclude that a total of 338,800 g of flour can be produced in the school garden. One loaf contains between 200 and 500 g of flour, so the teacher suggests assuming 338.8 g per loaf. Together they conclude that 1000 loafs of bread can be produced in the school garden. As this is obviously an unrealistically large number, the teacher adds: we probably miscalculated somewhere. This teacher remark marked the end of the activity.

8.3.3 *Reflection*

The 'grain and bread' activity is an example of how mathematics and science can be integrated in teaching. The context provides opportunities to explore biological aspects of grain growing and stimulates mathematical modelling and schematizing. In doing so, students learn how grain is used to make bread. The context of growing grain in the school garden also helps them mathematize the situation as multiplication (or repeated addition).

The teacher—who also designed the activity—saw how the context allowed students to engage with two STEM domains: science (biology) and mathematics. In supporting students' mathematizing process, she helped students find the multiplicative structure in the situation. She, however, translated this into calculating the answer, where in the end the number calculated proved to be wrong. In fact the focus on calculation moved students' attention away from mathematizing and from wondering if the answer could be correct. These processes are both key to integrating mathematics with STEM and are somewhat neglected here. Focus on round numbers or using a calculator might keep the student fixed on the mathematizing. Generally, our experiences here confirm the need for teachers' domain specific content knowledge and corresponding pedagogical content knowledge. Moreover teachers need to be able to scaffold these domain specific requirements as well as changing perspective between domains during the activity.

8.4 The Challenge of 'Maker Literacy' and 'Data Literacy'

8.4.1 Introduction

The continuous development of new digital artifacts influences education, work and leisure. The rapid pace of development poses challenges for carefully rethinking education and preparing for this changing technological context. In this section we describe two areas where tools and technologies are used and associated skills are needed. New teaching standards in the Netherlands require that students and teachers develop new 'literacies' in these areas (Kampman et al., 2024). Here, students and teachers need support from well-designed curricula and textbooks. We report here on two areas that were the focus of two different PLNs:

- Maker literacy, where the focus is working with the combination of hands, head and heart, using both simple technologies like hammer and nail and also high-tech tools like 3D-printing and working with programming languages like Scratch.
- Data literacy, which for example deals with artificial intelligence, vigilance on fake news, and computational thinking.

8.4.2 Maker Literacy

In primary education playing with and manipulating concrete materials is rapidly replaced by mere cognitive tasks. When children grow older, handwork is less valued by textbook authors and teachers. Although manipulatives are important tools for learning, this mode of learning is increasingly neglected in primary education. The overemphasis on cognition does not provide a proper balance for children to develop all of the STEM skills. The 'Maker Movement' or 'Maker Culture' emphasizes the use of hands and creation: learning-through-doing. This approach is supported by a variety of tools (Libow Martinez & Stager, 2014; Martin, 2015; Pijls et al., 2022).

The PLN that acted on this issue worked with different classroom examples (age group 11–12):

- How to build a scale model of a house, where you connect the manual work with the mathematical activity to get the right proportions.
- How to design a fantasy car where the requirements are: The car must be fast, four people must fit in the car, and the design must have an innovative look.
- How to use a 3D printer with children. This task was piloted at one of the PLN schools. The teacher provided support to accommodate the use of this technology in her teaching. However, the PLN discussion made clear this was an exception. Going forward, for this PLN participants planned to set up a collaboration between the primary school and a school for vocational education nearby.

8.4.3 Data Literacy

Digital technology and the internet have now made data collection and data sharing omnipresent, and data analysis and big data are common ingredients in the news, business, government and other social sectors. Being data literate and understanding digital technology surrounding us is thus increasingly important (ACME, 2023). Data literacy is an especially urgent issue when data and facts are routinely published with no checks or compliance with norms. Users of new technologies often find it difficult to understand how the technology has been used, according to which rules, and what data underpins it.

Against this backdrop, new national key learning objectives on digital/data literacy were recently developed in the Netherlands (Kampman et al., 2024). In line with this development a PLN consisting of primary school teachers, teacher educators and others developed different classroom tasks and activities to support data literacy:

- Makey Makey is an easy-to-use programming environment (Note: See https://makeymakey.com/). In one of the activities students (aged 8–10) used Makey Makey to design a prototype program code. Students were then encouraged to reflect on their work and critically analyze the 'computational' activity of making something "work".
- Using the Scratch programming environment for students between 10 and 15 years, led to a discussion in the PLN of the time needed for computational thinking activities, like coding and programming. Teachers also discussed various tools that they used. This discussion touched upon whether teaching and learning data literacy should be kept separate or included in other school disciplines like mathematics or language (Note: For background information see scratchjr.org and scratch.mit.edu.)

Two other topics for discussion were typical in the PLNs: namely 'critical thinking' and the use of artificial intelligence (AI). Critical thinking is needed for interpreting news and social media. Teachers in the PLN elaborated on ways to support students in participating productively in social media discussions. AI was discussed in one of the PLNs in relation to technology lessons, as a part of tools like Adobe Photoshop and Illustrator.

8.5 Discussion and Conclusion

8.5.1 Strong Pedagogy and Content Knowledge for Teaching STEM

Shulman defined Pedagogical Content Knowledge (PCK) almost forty years ago and his research (Shulman, 1986) still helps the understanding of domain specific pedagogy. In PCK content knowledge and pedagogical knowledge coincide. PCK represents the blending of content and pedagogy into an understanding of how particular aspects of subject matter are organized, adapted, and represented for teaching. This is especially true for STEM education (STEM-PCK) where teachers need both strong pedagogical and content knowledge and skills to be effective. Researchers have reiterated this for both science and mathematics (Gresnigt et al., 2014; Grgurina et al., 2014; Loughran, 2004).

8.5.2 Developing STEM-PCK

All the work from PLNs described in this chapter focused on the connection between pedagogy and content knowledge for the STEM disciplines. In our experience, working in PLNs resulted in the growth of teachers' skills and experience. However, teachers in the PLNs are not representative for all teachers. The PLN teachers were more interested in STEM subjects and their content knowledge exceeded that of many other teachers.

We observed that PLNs can play an innovative role within and across institutions. They help strengthen STEM activities in schools, building practical examples which can then be used by others, thus supporting the network of schools and teacher education institutes.

This relates to other learning for supporting the development of teachers' content knowledge and STEM skills and attitudes: teacher education curricula should incorporate a firm foundation of STEM domains. Additionally, adequate attention for STEM-PCK is more promising when there is someone within the school responsible for supporting STEM. This idea is not new. The 'ICT-coordinator' and 'mathematics coordinator' were introduced more than 25 years ago in Dutch primary education. However, these professionals focus on one STEM domain only. It would be helpful to organize STEM-wide professional support in the schools.

8.5.3 Recommendations

This chapter elaborated on the findings from the Dutch SPRONG STEM project (2021–2024). We discussed examples of how PLNs in different regions facilitated primary education STEM development. We described STEM education in its societal context and specific issues like alignment, integration, assessment, teacher education and inquiry-based learning. We have showed how the PLNs catalyzed regional and nationwide STEM education development. Drawing on these experiences, we make three general recommendations for developing STEM in primary education.

Recommendation 1—Design a connected STEM pedagogical content knowledge (PCK) foundation for all teachers involved

Developing teachers' content knowledge and their pedagogical content knowledge can help teachers realize high quality science (STE) and mathematics (M) activities in their teaching. This applies to individual STEM domains as well as STEM as an interconnected whole. We found that PLNs in which teachers, researchers and educators work cooperatively on themes adopted from a STEM perspective are a promising means to achieve this.

Recommendation 2—Involve multiple institutions and backgrounds in PLNs

As PLNs may provide a way to enhance STEM education, carefully composing PLNs is crucial. A variety of professional backgrounds in a PLN is essential. We found that in order to support STEM education, the PLN should include teachers, educators, researchers, and curriculum developers. Moreover, successfully learning and developing in a PLN depends on a broader regional approach, where PLNs reflect cooperation between school boards and higher education institutions.

Recommendation 3—Make 'STEM-connections' in national standards, and in curricula in primary education and teacher education

STEM education development grows out of local initiatives, where PLNs provide ready-made materials and ideas for practice. But more is needed. STEM education needs to be secured at the regional and national levels. Teacher education needs to explicitly embed STEM in the curriculum. Teacher education institutes can thus set an example for primary education. Moreover, both primary education and teacher education curricula are helped when the STEM domains and the interconnected nature of STEM are grounded in national standards.

8.6 Final Remarks

We began this chapter by identifying three challenges in STEM-education in Dutch primary schools. We focused on assessment and showed the importance of recognizing the complexity in the nature of the STEM domains. Exploring an activity in primary education, we elaborated on the M (for mathematics) in STEM. We found that integrating mathematics and science is hindered by a mere focus on mathematical procedures. These findings relate to the third challenge, namely teachers' PCK (pedagogical content knowledge). We saw that teachers' content knowledge is crucial for STEM in primary education, as it is essential for PCK development.

These challenges originated in recent developments in STEM education. PLNs consisting of teachers, educators and researchers, took broad and recognized issues in STEM education as the starting point to learn co-operatively. This cooperative learning subsequently formed the basis for the recommendations we make here. We take the position that developing STEM education in primary school requires a permanent dialogue between people in the field of primary education and researchers. In doing so we, researchers in the field, consider teachers as co-researchers, and aim to co-operatively develop STEM in Dutch primary education.

Acknowledgements We thank Elwin Savelsbergh for his helpful comments on a first draft of this paper. We thank Madhuvanti Anantharajan for her valuable text suggestions. Further, we thank all teachers and schools involved in this project. We especially thank Kim Baan, Mohamed Bakouri, Marit Benes, Marijke de Bruin, Geertje Damstra, Wendel Dulack, Petra Farber, Saskia de Jong, Linda Kaspers, Martin Klein Tank, Marit Kruiskamp, Gäby van der Linde Meijerink, Manon Nieuwenkamp, Corine Noordink, Jessica Okkerman, Esther Okoh-Van der Woude, and Miranda Wedekind.

References

ACME. (2023). *A new approach to mathematics and data education.* The Royal Society's Advisory Committee on Mathematics Education.

Bakker, M., Keijzer, R., & Hotze, A. (2023). Brood geeft aanleiding tot… [Bread leads to …]. *Terugkoppeling, 32*(4), 19–23.

Bom, P. L., Koopman, M., & Bijaard, D. (2019). Student teachers' use of data feedback for improving their teaching skills in science and technology in primary education. *European Journal of STEM Education, 4*(1), 09. https://doi.org/10.20897/ejsteme/6285

Crismond, D. P., & Adams, R. S. (2012). The informed design teaching and learning matrix. *Journal of Engineering Education, 101*(4), 738–797.

Davis, E. A., Petish, D., & Smithey, J. (2006). Challenges new science teachers face. *Review of Educational Research, 76*(4), 607–651.

Djoyoadhiningrat-Hol, K., & Klein Tank, M. (2023). *Trendanalyse Wetenschap & Technologie. Ontwikkelingen en uitdagingen bij Wetenschap & Technologie in het basisonderwijs* [Trend analysis Science & Technology. Developments and challenges in Science & Technology in primary education]. SLO.

English, L. (2015). STEM: Challenges and opportunities for mathematics education. In T. Muir, J. Wells, & K. Beswick (Eds.), *Proceedings of the 39th meeting of the international group for the psychology of mathematics education, PME 39* (Vol. 1, pp. 4–18). IGPME—The International Group for the Psychology of Mathematics Education.

Gal, I. (2024a). Adult education in mathematics and numeracy: A scoping review. *ZDM*. https://doi.org/10.1007/s11858-024-01549-z

Gal, I. (2024b). What do citizens need to know about real world statistical models, and the teaching of data modeling. In S. Podworny, D. Frischemeier, M. Dvir, & D. Ben-Zvi (Eds.), *Reasoning with data models and modeling in the big data era* (pp. 91–100). Minerva Foundation and Paderborn University. https://doi.org/10.17619/UNIPB/1-1815

Gravemeijer, K., Stephan, M., Julie, C., Lin, F.-L., & Ohtani, M. (2017). What mathematics education may prepare students for the society of the future? *International Journal of Science and Mathematics Education, 15*(Suppl 1), 105–123. https://doi.org/10.1007/s10763-017-9814-6

Gresnigt, H. L. (2018). *Integrated curricula: An approach to strengthen science & technology in primary education.* TU Eindhoven.

Gresnigt, H. L., Taconis, R., Van Keulen, H., Gravemeijer, K., & Baartman, L. (2014). Promoting science and technology in primary education: A review of integrated curricula. *Studies in Science Education, 50*(1), 47–84.

Grgurina, N., Barendsen, E., & Zwaneveld, B. (2014). Computational thinking skills in Dutch secondary education: Exploring pedagogical content knowledge. In *Proceedings of the 14th Koli Calling international conference on computing education research* (pp. 173–174).

Guérin, L. (2018). *Group problem solving as citizenship education: Mainstream idea of participation revisited.* Utrecht University.

Harlen, W. (1999). Purposes and procedures for assessing science process skills. *Assessment in Education: Principles, Policy & Practice, 6*(1), 129–144.

Hoogland, K. (2023). The changing nature of basic. *Frontiers in Education.* https://doi.org/10.3389/feduc.2023.1293754

Hotze, A. C., & Keijzer, R. (2017). Samenhang tussen rekenen-wiskunde en wetenschap en technologie [Connecting mathematics and science in education]. *Volgens Bartjens–ontwikkeling en onderzoek, 36*(5), 41–51.

Kampman, L., Driebergen, M., Oberink, M., Grgurina, N., Spronk, J., De Vries, H., & Klein Tank, M. (2024). *Conceptkerndoelen burgerschap en digitale geletterdheid* [Concept standards for citizenship and digital literacy]. SLO. https://slo-kerndoelen.files.svdcdn.com/production/uploads/assets/updates/DEF_kerndoelenboekje_BU_DG.pdf

Kirschner, P. A., & Stoyanov, S. (2020). Educating youth for nonexistent/not yet existing professions. *Educational Policy, 34*(3), 477–517. https://doi.org/10.1177/0895904818802086

Kneepkens, B., Van der Schoot, F., & Hemker, B. (2011). *Balans van het natuurkunde-en techniekonderwijs aan het einde van de basisschool 4* [Overview physics and technology education end of primary education]. Cito.

Laugksch, R. C. (1999). Scientific literacy: A conceptual overview. *Science Education, 84*, 71–94.

Libow Martinez, S., & Stager, G. (2014). *Invent to learn. Making, tinkering, and engineering in the classroom.* Constructing Modern Knowledge Press.

Lonning, R. A., & DeFranco, T. C. (2010). Integration of science and mathematics: A theoretical model. *Science and Mathematics, 97*(4), 212–215. https://doi.org/10.1111/j.1949-8594.1997.tb17369.x

Loughran, J. (2004). In search of pedagogical content knowledge in science: Developing ways of articulating and documenting professional practice. *Journal of Research in Science Teaching, 41*(4), 370–391. https://doi.org/10.1002/tea.20007

Maass, K., Geiger, V., Ariza, M. R., & Goos, M. (2019). The role of mathematics in interdisciplinary STEM education. *ZDM Mathematics Education, 51*, 869–884. https://doi.org/10.1007/s11858-019-01100-5

Martin, R. (2015). The promise of the maker movement for education. *Journal of Pre-College Engineering Education Research (J-PEER), 51*(1 (Article 4)).

Mulligan, J., & English, L. (2014). Developing young students' meta-representational competence. In J. Anderson, M. Cavanagh, & A. E. Prescott (Eds.), *Curriculum in focus: Research guided practice (Proceedings of the 37th annual conference of the Mathematics Education Research Group of Australasia)* (pp. 493–500). MERGA.

Oonk, W., Keijzer, R., & Zanten, M. A. (2020). Integration of mathematics and didactics in primary school teacher education in the Netherlands. In M. Van den Heuvel-Panhuizen (Ed.), *National reflections on the Netherlands didactics of mathematics* (ICME-13 Monographs ed., pp. 121–146). Springer. https://doi.org/10.1007/978-3-030-33824-4_8

Pedastre, M., Maeots, M., Siiman, L. A., De Jong, T., Van Riesen, S. A., Kamp, E. T., Manoli, C. C., Zacharia, Z. C., & Tsourlidaki, E. (2015). Phases of inquiry-based learning: Definitions and the inquiry cycle. *Educational Research Review, 14*, 47–61.

Pijls, M., Van Eijck, T., Kragten, M., & Bredeweg, B. (2022). Activities and experiences of children and makerspace coaches during after school and school programs in a public library makerspace. *Journal for STEM Education Research, 5*, 1–24.

Post, T. (2019). *Fostering inquiry-based pedagogy in primary school: A longitudinal study into the effects of a two-year school improvement project.* TU Twente.

Prenger, J., Pleumeekers, J., Van Zanten, M., Schmidt, V., Teunis, B., Brand, M., & Bron, J. (2023). *Conceptkerndoelen Nederlands en rekenen en wiskunde* [Concept standards Dutch language and mathematics]. SLO.

Rohaan, E. J. (2009). *Testing teacher knowledge for technology teaching in primary schools.* TU/e. https://doi.org/10.6100/IR653226

Shulman, L. S. (1986). Those who understand: Knowledge growth in teaching. *Educational Researcher, 15*(2), 4–14.

SLO. (2006). *Kerndoelen primair onderwijs* [Key objectives primary education]. OCW.

Van Aalderen-Smeets, S., & Walma van der Molen, J. (2013). Measuring primary teachers' Attitudes toward teaching science: Development of the dimensions of attitude toward science (DAS) instrument. *International Journal of Science Education, 35*(4), 577–600. https://doi.org/10.1080/09500693.2012.755576

Van den Heuvel-Panhuizen, M. (Ed.). (2020). *National reflections on the Netherlands didactics of mathematics: Teaching and learning in the context of realistic mathematics education* (ICME-13 Monographs ed.). Springer. https://doi.org/10.1007/978-3-030-33824-4

Van der Aalsvoort, G., Van der Zee, S., & De Wit, T. (2020). Improving science skills by practicing geometry and measurement in Kindergarten. *Early Child Development and Care, 190*(4), 537–548. https://doi.org/10.1080/03004430.2018.1482889

Van Graft, M., & Klein Tank, M. (2018). *Wetenschap & technologie in het basis-en speciaal onderwijs. Een richtinggevend leerplankader bij het leergebied 'Oriëntatie op jezelf en de wereld'* [Science & technology in primary and special education. A guiding curriculum framework for the learning area 'Orientation on yourself and the world']. SLO.

Van Keulen, H., & Slot, E. (2013). *Vaardigheden Rubrics Onderzoekend en Ontwerpend leren* (pp. 36). School aan Zet. https://www.fi.uu.nl/publicaties/literatuur/2013_vroo.pdf

Van Keulen, H., & Slot, E. (n.d.). *Excellentiebevordering door middel van onderzoekend en ontwerpend leren. Vaardigheden rubrics onderzoeken en ontwerpen (VROO)* [Promoting excellence through research and design learning. Researching and designing skills rubrics (VROO).]. School aan Zet. https://ktwt.nl/wp-content/uploads/sites/40/2015/08/vaardigheden_rubrics_onderzoeken_en_ontwerpen_vroo_.pdf

Voogt, J., & Pareja Roblin, N. (2012). A comparative analysis of international frameworks for 21st century competences: Implications for national curriculum policies. *Journal of Curriculum Studies, 44*(3), 299–321. https://doi.org/10.1080/00220272.2012.668938

Chapter 9
Inspiring Teachers: The Story of the STEM Teacher Enrichment Academy Primary Program

Jennifer Way, Katherin Cartwright, Christine Preston, and Vilma Galstaun

Abstract The purpose of this chapter is to demonstrate how the common experience of an evidence-based professional learning program can inspire Australian primary teachers from a range of schools to engage their students in diverse, innovative interdisciplinary STEM projects, unique to their own local contexts. In the period 2017–2024, the STEM Teacher Enrichment Academy: Primary Program worked with approximately 500 teachers from 147 schools. We propose that the combination of research-based education goals, an aspirational approach to supporting change in teacher pedagogy and the progressive scaffolding of both teacher and student learning has resulted in a successful STEM education program.

Keywords STEM integration · Primary education · Professional learning

9.1 Introduction

The story of the *University of Sydney STEM Teacher Enrichment Academy* (Australia) begins with serendipity. In 2014, three 'forces' coincided to create the *University of Sydney STEM Teacher Enrichment Academy* (hereafter 'STEM Academy'). (1) Political and education agendas in Australia and overseas began promoting STEM education as being necessary for prosperous economies and for addressing global

J. Way (✉) · K. Cartwright · C. Preston · V. Galstaun
University of Sydney, Sydney, Australia
e-mail: jennifer.way@sydney.edu.au

K. Cartwright
e-mail: katherin.cartwright@sydney.edu.au

C. Preston
e-mail: christine.preston@sydney.edu.au

V. Galstaun
e-mail: vilma.galstaun@sydney.edu.au

S. Earle et al. (eds.), *Primary Science Learning for Children, Teachers, and Communities*, https://doi.org/10.1007/978-981-95-0561-6_9

issues (Honey et al., 2014; Office of the Chief Scientist, 2014; Prinsley & Johnston, 2015); (2) The university was engaged in nationally funded projects to inspire cross-faculty collaboration to promote mathematics and science education, and enhance teacher education (IMSITE project, Goos et al., 2017; Goos & Bennison, 2018; ETMST project, Maconachie & McInnis, 2017); (3) A generous benefactor offered the university the funds to establish a school outreach program, to be a joint venture between the Faculties of Education, Science, and Engineering and Information Technology. The STEM Academy was born—first with a secondary school program, then with the primary school program.

This chapter is written by founding team members of the Primary Program and in it we share how we developed our own practice as teacher-educators working in the interdisciplinary space over 8 years. Each team member is a full-time academic, specialising in our own discipline—mathematics, science, technology—for research and lecturing. As the approach taken by our program has not been previously fully documented, writing this chapter has been an enlightening reflective process for the team. Therefore, the reader will find the chapter scattered with reflective comments to indicate our perspectives. Learning from each other, from our collaborations with practicing teachers participating in the professional learning and from the evaluative research in the early years of the initiative, the program has evolved into a resilient and highly successful entity. We inspire the teachers, and the teachers inspire us.

In the next section we provide an overview of the professional learning model. The subsequent section elaborates the design of the model by detailing what we believe to be the two key ingredients for the success of the program—the aspiration of integrated STEM projects and our tri-level approach to scaffold teacher and student development. To highlight the diverse ways in which schools respond to the program by creating projects unique to their own school communities, the fourth section describes three examples of end-of-program project outcomes. We conclude with recommendations about the aspects of the program which may be transferable to other contexts.

9.2 The Program Model

9.2.1 The Participants

Two primary programs for in-service teachers are run in parallel each year, one in the city-metropolitan area and one in a regional centre, such as a large country town. Entry to the program is via an expression of interest that must be submitted by the school Principal, with the expectation that there is a whole-school commitment to developing STEM education. The Sydney-based program often fills the available places from a waiting list, but invitations are sent out to surrounding local schools for the regional program. The programs are inclusive of all education systems, so the cohorts are usually composed of a mixture of state schools, catholic schools

Table 9.1 School numbers across the program 2017–2024

Year of program	Number of schools			
	State	Catholic	Independent	Total
2017 Regional only (Wagga Wagga)	8	4	1	13
2018 Regional (Coffs Harbour)	11	1	3	15
2018 City/metro	14			14
2019 Regional (Orange)	8	3	3	14
2019 City/metro	10	2	3	15
2020/2021* Regional (Newcastle)	4	3		7
2020/2021 City/metro	8		4	12
2022 Metro only	4		6	10
2023 Regional (Lismore)	6	5	1	12
2023 City/metro	5	4	2	11
2024 Regional (South Coast)	9	1	2	12
2024 City/metro	9		3	12
Total	**96**	**23**	**28**	**147**

* 2020 and 2021 are presented together as the schools remained in the program into 2021 due to COVID pandemic interruptions in 2020

and independent schools (see Table 9.1) and cover a range of demographics and socio-economic communities. The Independent schools include both secular and faith-based schools.

Each program accepts 10–15 schools, and the funding model supports four teachers from each school to be released from teaching duties to attend the five professional learning days. The program targets teachers of Grade 3 and 4 classes (Students aged 9–10 years), though some schools include a specialist teacher such as the librarian or a technology-support teacher in the team. Our focus is on the middle years of primary because it allows time for the students to build their STEM capabilities over several years. Small country schools might only send two teachers or collaborate with another small school to create a team. Principals and curriculum representatives from each school sector receive a standing invitation to attend any of the sessions, so each Academy has several observers. By 2024, the STEM Academy Primary program had worked with approximately 500 teachers from 147 schools across the state of New South Wales.

9.2.2 *Program Design*

The structure of the program consists of two full days of professional learning at the start of the school year, another two days mid-year and a final day at the end of the year. In between each set of days, the teachers meet online for an after-school networking session, and also have an in-school visit from their assigned mentor. Table 9.2 provides an overview of the structure and organisation of the content of the program. The STEM Academy is a teacher professional learning program accredited with the NSW Education Standards Authority, meaning that teachers can claim participation towards maintaining or advancing their teacher registration and accomplishment level.

The overall professional learning design is guided by Desimone's (2009) framework for 'studying' professional development which was derived from the extensive literature on effective approaches for teachers' professional learning. A summary of how our program addresses the five components of the framework (Desimone, 2009) follows:

(a) *Content focus*: A substantial portion of the program focuses on developing teacher knowledge of STEM education processes, some specific disciplinary content and how their students are likely to learn the content and develop the skills, and how to access expertise they do not possess. As each team member has expertise in a different discipline, we co-design each major activity for the teachers, to create a balance between supporting discipline-specific knowledge and ensuring that we model interdisciplinary thinking.

Table 9.2 STEM Academy Primary program design

Event	Purpose and main activity
Pre-session preparation	Each team considers the school's and their own individual positioning on the *STEM Integration* and *STEM Project Pedagogy* continuums
Days 1 & 2	Research base for integrated STEM education and introduction to the Academy's tri-level approach to STEM education Role of technologies in STEM, and potential issues in the Engineering design-processes Team planning of first actions in schools
Between	Online session: inter-school networking with a focus on teaching experiment successes, difficulties and resources Mentor supports the school teams in their implementation and further planning. Teachers continue with cycles of plan-implement-reflect
Days 3 & 4	Design-thinking and integrated STEM project structures. (e.g. Stanford Design Thinking model) Pedagogy of integrated approaches, including assessment of student learning Teachers prepare a timeline for commencing an integrated STEM project, a staff and community engagement plan, an assessment and evaluation plan, and a list of anticipated challenges and possible solution strategies
Between	Online session to support community networking and access to external resources Mentor supports the school teams in their implementation and further planning. Teachers continue with cycles of plan-implement-reflect
Day 5	Showcase day. Input from inspirational speakers. Teachers present their journeys and their individual and collective achievements

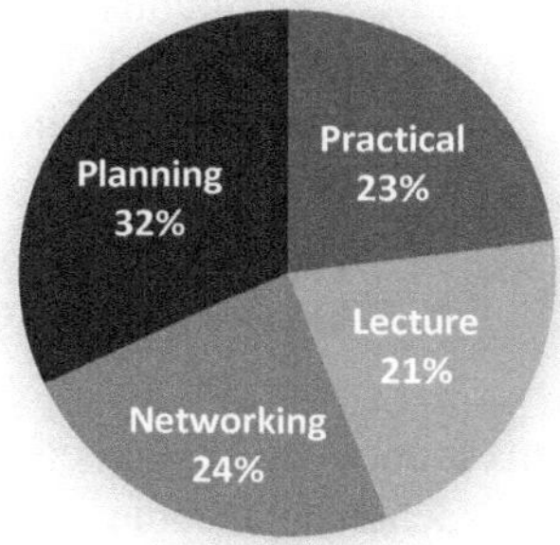

Fig. 9.1 Proportion of professional learning activity types

(b) *Active learning*: A variety of activity types is recommended. The program is composed of a balance of four main types of activity conducted during the contact time with teachers: practical workshops and reflective discussions, short lecture-style presentations, scaffolded networking sessions including short reports from the schools, and planning and discussion time for school teams (Fig. 9.1).

(c) *Coherence*: Coherence has two main components, consistency with teacher knowledge and beliefs, and consistency with state or school system polices and reforms. This aspect of the professional learning is a key challenge for the program. Although the *National STEM School Education Strategy, 2016–2026* (National Council, 2015) calls on educators to promote STEM education, and STEM initiatives in primary schools are encouraged by the state government Department of Education, there is no actual STEM syllabus to guide schools and the dominant thinking in schools tends to be about separate disciplines. The STEM Academy's approach to STEM education requires some level of inquiry pedagogy. We have found that most teachers believe such an approach can be beneficial to student learning, but they experience a range of barriers to implementing inquiry projects (Way et al., 2022). Therefore, the major focus of the Academy is to challenge teachers to adjust both their pedagogy and the organisational structures in their schools, and to provide support for achieving these goals.

(d) *Duration*: The recommendation for supporting teachers in pedagogical and intellectual change (Desimone, 2009) is about 20 hours of contact time spread across a semester. The STEM Academy operates across a full year, with 25 hours of direct work with teachers, an additional 5 hours in the end-of-program Showcase Day, and several hours of mentoring from the team members, often including a visit to the school.

(e) *Collective participation*: Collaboration between teachers has been identified as critical for supporting teacher learning and changes in practice. The STEM Academy scaffolds both within-school collaboration and inter-school collaboration. We recruit school teams of two to four teachers and encourage them to collaborate in planning and implementation. We also endeavour to nurture a community of learning through networking mechanisms such as mini-presentations by schools, mixing of school-team members in some practical activities, and the sharing of ideas and resources via online platforms.

9.3 Distinctive Features of the Program

In this section we explain our practice in terms of how we situate, support and challenge teachers to become creative and responsive within their unique school contexts as they plan and implement localised, authentic STEM teaching and learning experiences. From the inception of the STEM Academy, various forms of data have been collected to evaluate or investigate aspects of the program, with a focus on monitoring change in teachers' practice and change in student engagement and learning. Some of the data have been reported directly to the Academy's Academic Committee and Advisory Board, and some has been published in research papers (e.g., Anderson et al., 2019; Way et al., 2022). Data is also reported directly back to the schools as a form of feedback on the impact of their efforts. Data sources have included teacher and student surveys, school case studies and teacher interviews. Our claims about the success of the program are grounded in this data and supplemented by our own observations.

The goals of the STEM Academy are aspirational in the sense that teachers are expected to, across the period of one year, effect some change in their perception of STEM education, and their pedagogy, and practice. The anticipated changes are framed by two continuums operating in the context of STEM projects. One continuum that maps progress from a separated disciplinary curriculum approach towards a transdisciplinary approach was adapted by the primary team from Vasquez (2015) (Fig. 9.2).

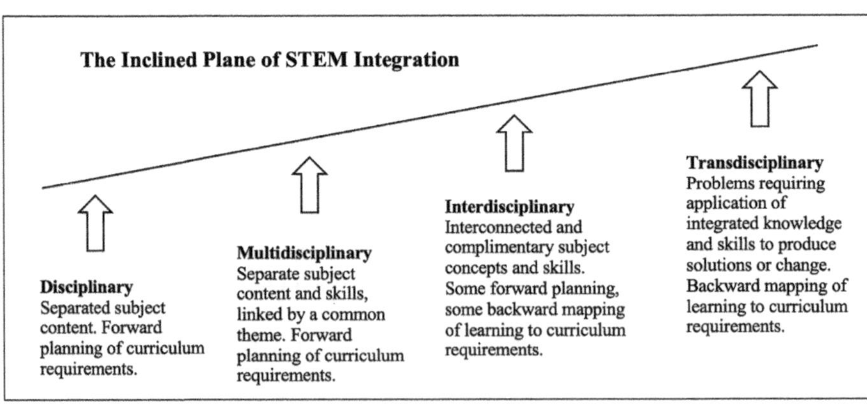

Fig. 9.2 Continuum of STEM integration adapted from Vasquez (2015, p. 13)

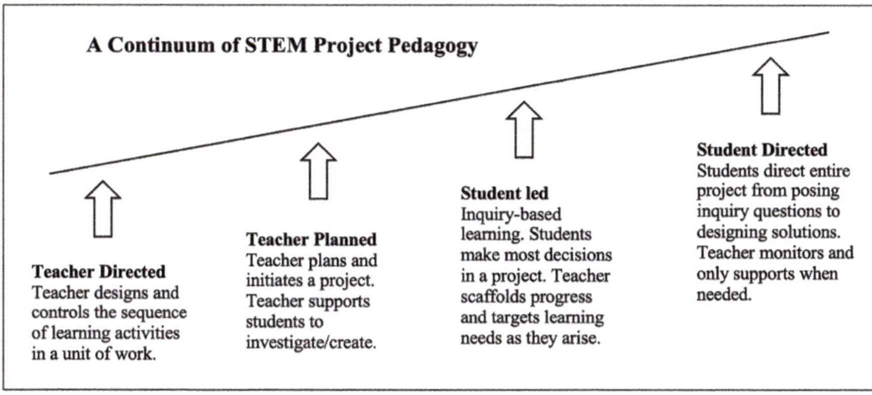

Fig. 9.3 Continuum of pedagogy in STEM projects

The other continuum developed by the STEM Academy Primary team, describes a corresponding shift away from teacher-directed pedagogy towards student-directed pedagogy (Anderson et al., 2019) (Fig. 9.3).

At the beginning of the STEM Academy Primary program each teacher is invited to 'locate' themselves on each continuum and consider their own aspirations for movement 'upwards', whether it be a small or large change. School teams are also encouraged to consider their collective goals in relation to their perceived positioning of the school staff in general. This exercise highlights to the teachers that they should judge their success in relation to their own progress rather than in comparison to other schools who may have 'higher' start points.

One measure of the success of the program is teacher change related to these two continua. Surveys conducted with the 2018 and 2019 cohorts (n = 152) revealed the top four changes in teacher perceptions towards STEM teaching and learning, accounting for 74% of the teachers, to be: "Acquired increased knowledge in STEM" (55); "Importance of subject integration" (24); "More confidence as STEM teacher" (20); "Student-led learning is achievable" (13) (Tully, 2020, p. 10). These same cohorts of teachers reported changes in their practices such as: "Provide more opportunities for student leading" (58); "Apply design process" (34); "More integration across subjects" (23) (Tully, 2020, p. 12).

We credit much of the influence on teachers to two distinctive features of the Primary program, each evolving from two very different sources. The first feature is the definition of the aspiration of the STEM Academy which is derived from previous research findings and communicated to the teachers via the desired characteristics of STEM project. The second feature is the tri-level approach to developing STEM capabilities which emerged from the team's interaction with teachers in first two years of the program.

9.3.1 Research-Based Characteristics of Quality STEM Projects

When developing the original design of the program during 2016, we drew on the available research on STEM education in the primary school context to derive six characteristics of STEM projects likely to maximise positive learning outcomes for students, and we have continued to assess them against subsequent research findings (e.g., Corrigan, 2020). We use the term 'project' broadly to mean a sustained focus on achieving an identified goal, which allows teachers to adapt other approaches being used in their schools, such as problem-based learning, inquiry-based learning or project-based learning (Lowrie et al., 2017). Achievement of the six characteristics sets the aspirational goal of the Primary program and the teachers are encouraged to attempt such a project by the end of the year.

1. **Real problems, real solutions**: *Finding solutions to a meaningful problem, question or challenge, situated in a real-life context relevant to, or of special interest to, the current cohort of students.* The benefits of authentic learning contexts arise from the meaningful engagement of students with something that matters to them, and the sense of empowerment they experience when they realise that others will listen to their concerns and ideas if they put the effort into the investigation processes (Lowrie et al., 2017; Rosicka, 2016; Sanders, 2012). This characteristic represents a critical decision made by the Academy to *not provide* units of work but rather support teachers to develop their own locally situated projects.
2. **Student-centred inquiry** *processes including hypothesising, planning, designing, creating, testing, decision-making, redesigning, and critically evaluating.* Active learning approaches in which students take some responsibility for their own learning maximise the benefits of engaging in authentic tasks, including the development of general capabilities, such as collaboration skills (Corrigan, 2020; Hudson et al., 2015; Rosicka, 2016). This characteristic reflects the 'continuum of pedagogy in STEM projects' mentioned in the previous section and is a major focus of support during the professional learning sessions.
3. **Integration**: *Genuine integration of the disciplinary characteristics of science, physical technology and engineering, digital technology, and mathematics.* Such integration is achieved through focusing on processes—the ways of thinking and working that are common to the STEM disciplines, where knowledge (curriculum content) emerges from a 'need to know' basis, in order to enable progress towards the project goal (Becker & Park, 2011; Corrigan, 2020; English, 2016; Treacy & O'Donoghue, 2014).
4. **Creativity and evidence-based reasoning**: *Valuing students' ideas and promoting critical thinking based on data.* This characteristic is based on the premise that creative thinking and innovation are important ingredients in STEM projects, but do not develop into viable solutions without testing and evidence gathering. This highlights the central role played by design processes (Engineering) in STEM education (English & King, 2015; Estapa & Tank, 2017;

Hathcock et al., 2015). The development of 'design thinking' is emphasised in the STEM Academy, for example, creating a 'wind-powered car' prototype in the programme is an influential shared-experience for the teachers, and one which most teachers repeat with their own students (see Anderson et al., 2019; Preston, 2020).

5. **Communication**: *A variety of approaches to explaining and justifying processes and solutions.* Communication skills, including those utilising digital-technology modes, are needed to document individual and collective learning journeys, to share ideas and findings, to convince various audiences of 'solutions', and to celebrate achievements. As one of the '21st Century skills' crucial to STEM learning (Dare et al., 2021), communication permeates all aspects of students' engagement in STEM projects.
6. **Future aspirations**: *Raises awareness of STEM in society, culture and/or career pathways.* One of the underlying motivations for STEM education is to inspire young people to achieve higher levels of education and encourage them to undertake STEM related careers. This aspect of STEM education is particularly significant for under-represented groups in STEM, including females, Aboriginal people and students from disadvantaged backgrounds (Holmes et al., 2018; Lowrie et al., 2017). While we have little information about how teachers work towards achieving this characteristic, the STEM Academy 2023 student surveys indicate a shift in student knowledge and aspirations between the start of the year and the end. For example, we saw shifts from student statements like '*I like to work in maths because I'm good at maths*', to statements that reveal the influence of their recent experiences, like '*I would like to do jobs in engineering and IT because I have watched videos of Engineering. I hope to find my path to success. I have tried several Engineering skills and projects and hope to invent them when I am older*'; and '*If I was a biologist I would like solving problems and finding new solutions to problems*'. At the start of the year 30% of the students expressed no interest in a STEM career and at the end of the year this had reduced to 16%.

9.3.2 The Tri-level Approach to STEM Education

During the first two years of the program, it became apparent to the team that many of the teachers would benefit from additional scaffolding to effect change in their pedagogy, confidence and management of organisational constraints. Pedagogical challenges arise when teachers attempt to increase their use of student-led inquiry to open-up opportunities for the 21st Century learning skills of communication, collaboration, creativity, and critical thinking (Dare et al., 2021). Inspired by observations of teacher practices in our 2018 Academy cohort, the team innovated the 'tri-level approach' (see Table 9.3), which focuses on building the essential skills that students need to engage effectively with integrated STEM projects and scaffolds the pace of pedagogical change required by many teachers (Way et al., 2022).

Table 9.3 Overview of the Sydney STEM Academy Primary program tri-level approach

Level title	Purpose and Characteristics	Examples
Level 1 STEM skills	Short (e.g., 15 min), focused, repeatable tasks to develop students' basic skills and learning attitudes Can be embedded within any curriculum area or used as lesson breaks	Collaborative group work using clearly defined assigned roles Active-listening and careful explanation of ideas Quick construction challenges (learning from failure) Basic robot programming Collecting, displaying and interpreting data Observation and recording techniques, including representational drawing
Level 2 design process	Substantial tasks to develop plan/ design—make—test—redesign processes Usually embedded within the Science curriculum or other existing unit of work containing a suitable problem/design context	Designing a wind-powered car within a unit of work on forces Hypothesis testing and data-based decision-making when changing the wing-length of a paper airplane or helicopter Proposing creative solutions to a watering problem with plants, following information research
Level 3 STEM projects	Aspiring to the 6 project characteristics Sustained multi-disciplinary projects to solve authentic problems Involvement with the broader community and sources of expertise Focus on general capabilities like critical thinking, collaboration, communication, creativity Responsiveness to emerging needs for new knowledge or skills as the project evolves	Restoring an ecosystem like a pond or barren patch of land Developing more sustainable practices in the school like rubbish management and recycling Repurposing an area of the playground or a room to better meet the diverse needs of students Designing a way to honor the history and/or cultural heritage of the school or local area

At the end-of-year STEM Academy showcase, many of the school teams refer to the tri-level approach as being integral to the success of their projects and they use the three levels as a 'language' to communicate their progress. Often mentioned is the value of repeating Level 1 tasks to establish in their students the collaboration skills and the culture of 'learning from failure' that is needed for them to cope with the challenges of the design process of Level 2. Many teachers come to realise that skill-based achievements, such as learning to program a robot (Level 1), are actually tools to be applied later in more substantial design tasks (Level 2), and in authentic problem-solving tasks involving higher-order skills (Level 3). They report that students begin to see purpose in mastering fundamental lessons (like making accurate measurements in mathematics) because they need to apply the knowledge and skills in meaningful tasks later. Teachers see benefit in working through the levels sequentially, but also mention the flexibility of 'going back' to a previous level to respond to detected needs in their students. Some schools, for a variety of reasons, do not achieve the goal of a Level 3 STEM Project by the end of the year, but carry over their efforts to the next school year and build on the established skills of their students. Interviews conducted across five schools from the 2020/2021 STEM Academy cohort revealed that not only did the teachers find the tri-level approach effective in their own classrooms, but also that it served as a useful structure for involving other teachers and supporting a school-wide change (Way et al., 2022).

9.4 Examples of Level 3 STEM Projects

This section presents examples of the unique STEM projects that teachers and their students have created to meet the needs of their school communities. These illustrations of practice demonstrate the impact of the STEM Academy and demonstrates the great diversity of what teachers can achieve in their own contexts when supported throughout the course of one year to 'go their own way'. In some schools, the commitment by the teachers and school leadership to embrace STEM education in meaningful ways has resulted in a sustained school commitment in the years following participation in the STEM Academy.

At the end of the year-long program, schools present their 'STEM journey' at a showcase day, and those presentations are the source of data for this section of the chapter. Table 9.4 presents a thematic analysis of the 94 available showcase presentations for schools that attempted a Level 3 Project. As shown in Table 9.4, a wide variety of authentic topics were investigated. Many of the projects focused on exploring and finding solutions to issues within their school grounds or immediate school community. For example, the theme that arose most often was *Wellbeing/school learning environment*. Within this theme there are several foci including the creation of effective learning areas (inside and outside), play spaces that cater for the diverse needs of learners, and improvement to the play facilities accessed by students. *Environment/sustainability* was another major theme that threads through many of the projects. Diversity of projects in these cases related to the creation of sustainable gardens, investigating future sustainable practices, dealing with waste in school environments, and ways to support and protect surrounding flora and fauna. Numerous projects related to the environment, connected to 'country' and local Aboriginal culture. The remaining projects dealt closely with *Empathy* with the people within the school's community, regarding their current wellbeing or preparation and protection during natural disasters (which several Academy schools had experienced first-hand). These projects enabled students to think about solutions to problems outside of their own immediate needs, and the stories shared from these projects were often emotive in nature.

The final outcomes of Level 3 STEM Projects vary from being incomplete, but with accomplishments acknowledged (a few schools), a school-community event celebrating student learning and prototype solutions (many schools), to actually

Table 9.4 Themes from Level 3 STEM projects 2017–2023

Main theme	Focus		Number
Wellbeing—school learning environment	Multi-sensory wall/ space Playground improvement Calming space Outdoor learning space Indoor STEM space, classrooms	Bike storage Playground accessibility Canteen food delivery Handball courts	30
Environment—garden	Bush tucker garden Butterfly garden Edible outdoor garden	Community/local garden Cultural garden Yarning circle	15
Waste—recycling	School bins Rubbish issues		12
Sustainability—environment	Fauna e.g., koalas, bees Marine fauna	Outer space Flora and fauna	11
Sustainability	Energy Materials Food	Water Up-cycling	10
Empathy/wellbeing—local or school community-focused	Connecting to community Safety e.g., road Needs e.g., toys for	disadvantaged children low-tech no-tech toys	9
Empathy—disasters, environment	Floods Fires Drought		7

enacting solutions, such as planting the garden and supplying vegetables to the local soup-kitchen (some schools). We consider all of these scenarios to be successful project conclusions. A common outcome for almost all the projects is heightened parent and community recognition of the value of students engaging in this type of STEM education. Most schools also share plans for expanding STEM education across the school and continuing the embedding of STEM education into the school culture during the following year. To illustrate the 'enrichment' aspect of the STEM Academy, we offer an extract from a letter written by a parent of a boy (Max, pseudonym) attending a school from the 2019 cohort, located in a low socio-economic outer suburb of Sydney. The letter refers to the showcase of the STEM projects, held in the school hall, which attracted a record number of parents into the school grounds. According to the school principal, a major outcome of STEM Academy involvement was improved school-parent communication and increased parent interest in the children's learning.

> This year Max's curiosity and enthusiasm for learning has grown tremendously due to his teacher and STEM lessons. I feel Max is fully engaged because he is involved in his learning. He is motivated and excited to talk about what he has learned, built, coded, the experiments,

> what has worked out and not worked out as planned and how to modify...The STEM Showcase was a fabulous opportunity to see the outcome of the weeks of experimentation and collaboration I had been hearing about from my son. The pride, quality of work and eagerness of the students to show and explain their design experiment with a demonstration was exciting to see and a pleasure hear. I feel extremely proud and grateful for how far Max has come this year. I would like to see STEM continue and extend to all stages at school. Benefits I have seen go beyond knowledge of Science, Technology, Engineering and Math to develop skills in collaboration, teamwork, problem solving, critical thinking skills and emotion regulation.

Three examples of school projects have been selected to illustrate the contextualised nature of STEM projects. (A fourth example can be found in Preston et al., 2019). As mentors of these schools, the authors of this chapter feel privileged to have worked alongside schools in their STEM journeys. Each was chosen for the project's authenticity to their school/community context and have been discussed in relation to their alignment with the STEM academy's six characteristics of STEM projects.

9.4.1 Project Example 1: The 'Grandbuddy' Project

STEM Academy 2018

Driving question: How can we help enhance the life of the elderly at the nursing home, using technology?

Western School students had visited a local aged-care home to sing for the residents throughout the previous year. It was this prior relationship with their community that the school wanted to expand upon to make deeper connections to the residents. The teachers had a deep desire to make learning authentic, and a drive to embed social responsibility in their students through community engagement. Therefore, they decided to look at ways for students to go beyond entertaining the nursing home residents to help them with everyday real problems. The teachers reported spending a considerable length of time focused on the first step in the Stanford Design Process (Hasso Plattner Institute of Design, 2018)—*Empathise*. The students spent time building relationships with the residents: playing games, interviewing them about their life and making timelines, visiting residents weekly, and participating in lessons regarding aging, illness and dealing with death. The *Empathise* phase of the project was an opportunity for the teachers to embed aspects of the project across key learning areas such as English (involving writing, reading and oral communication), and Personal Development/Health. Although this relationship-building was time consuming, the teachers noticed benefits to the students' learning regarding personal development and skills in recounting personal histories. As the project moved into the *Define* and *Ideate* phases, students utilised what they had learnt about their 'grandbuddy' to listen to the residents' real problems and collaboratively brainstorm possible solutions. Some of the problems discussed included helping residents with their memory, feelings of loneliness, finding their glasses, keeping active, and feeling happy. During the project the teachers reported that the relationships that

Table 9.5 Driving questions and solutions created by groups of students

Driving question	Solution
How can we help: • *Tony with his feelings of loneliness?*	QR codes linked to videos
• *Phillipa's memory?*	Make a card game using Scratch with levels and prizes
• *John to be happier?*	Large interactive map of Egypt
• *Pat find her glasses?*	Glasses tracker
• *Nancy hold her jewellery?*	3D printed electronic jewellery box
• *Keep Allan's mind active?*	Magnetic Chess set
• *Nola with her depression?*	Write a 'choose your own adventure' story with stop motion videos

ensued were much deeper than they could have imagined, and this motivated the students to develop a personalised driving question and designing a solution (See Table 9.5).

Classroom lessons during the *Prototyping* and *Testing* phases were opportunities for genuine integration where students engaged in student-centred inquiry by designing, creating, and testing their ideas. Students developed their 'working scientifically' skills and explored technology aspects of their designs such as stop motion videos, robotics, 3D printing, and electric circuits.

The students met with the residents at each step of the process to show the 'grandbuddies' their progress and get feedback on their designs. Students often redesigned their solutions based on the feedback provided by their grandbuddy, who valued their creativity. This Level 3 project took over two school terms (close to 6 months) and was celebrated with a culminating showcase day at the nursing home where students explained their processes and solutions. The residents' family members attended along with the students, teachers and STEM Academy academics. The success of Western School's project also resulted in nomination as a finalist for the 2019 NSW Premier's Awards and they were invited to present on their project at the Social Ventures Australia (SVA) Thought Leadership Gathering. These nominations acknowledged the project for its connections with the community and outside agencies, and its impact on student learning.

9.4.2 Project Example 2: Safety Around Our School

STEM Academy 2022

Driving question: How can we make the area around our school safer?

Metro School is in a busy area with street access on three sides of the school. The streets are single lane and there is no dedicated drop off or pick up lane for students travelling by car. Parents are often double-parked, blocking traffic flow and students exit from or into a car in unsafe areas. The ongoing concern within the school community about the safety of students travelling to and from the school provided the teachers and the Grade 4 students with a real problem that needed to be addressed with real solutions. To better understand the problem, students surveyed the road traffic, the behaviour of parents driving, and the pedestrians (including students) in other classes.

They investigated the bottlenecks that made it unsafe for students, specific times in the day that were problems for students' safety, and what constituted unsafe behaviours. Students analysed their data and identified three major factors making travelling to school unsafe.

1. Parents were being unsafe dropping students off at the school entrance, often in the middle of the road, doubled-parked.
2. Students running across the road, not using the pedestrian crossing because parents had dropped them off on the opposite side of the street.
3. There was no 'pull-in' area for parents to drive in to 'kiss and drop' students.

Students researched what they could do to help make the road area safer. They asked questions about traffic accidents in the area, who was affected, and further consequences. They identified people in the school and the community who could provide them with information, such as the school road crossing attendant, the police, the local council, their parents, the principal and staff.

The teachers invited the duty road safety officer from the local police precinct and the school's road crossing attendant, to speak to the students at what we called a 'Wonder Day' where students asked prepared questions. After the Wonder Day, the students were more knowledgeable about road and pedestrian safety and were ready to devise solutions to their identified problems. Students explored various creative ways for presenting their solutions. Some groups opted for technology-based solutions using an amination app or a micro-bit on an iPad or laptop that involved using skills such as coding their road safety solution, other groups used Lego and movable construction materials. Some prototypes were posters for pedestrian safety, and others developed fact sheets using *Canva*. Students used photos from their investigations and information as evidence to explain and justify the processes and solutions they were proposing.

Students became aware that any changes to the roads and the installation of dedicated 'kiss and drop' zones were the responsibility of the local council. In response, they decided to write to the mayor about their concerns. After testing and retesting their prototypes, students were ready to present their solutions on a special showcase day, inviting the mayor, parents, local area police road officer, school road crossing attendant, the school principal, teachers and other classes to view their proposed solutions to the community-based problem. The mayor attended the showcase, listening to the students and viewing their possible solutions and announced that their concerns were to be discussed at the next council meeting. The overall impact of the STEM project was to raise community awareness of student safety around school and also the recognition of the value of STEM education.

9.4.3 Project Example 3: Sustainability Spotlight on School Litter

STEM Academy 2023

Driving question: How can we reduce litter at Seaside School?

The school implemented various sustainability initiatives including a plastic-free canteen where students bring their own cutlery, waste free Wednesdays, the Planeteer leadership programme and the Stephanie Alexander kitchen garden programme (The Stephanie Alexander Kitchen Garden Foundation, 2014). Despite these efforts, the Grade 3 and 4 students identified litter in the playground as a significant issue, especially considering the proximity to the ocean.

To address the school's waste problem, students analysed the primary locations of litter and collected data to identify the main causes. Students identified four contributing factors: school fences allowed rubbish in and out, birds were opening children's lunchboxes, poorly designed bins, and students' behaviour of littering and not taking responsibility for their rubbish. Six Grade 3 and 4 classes (aged 8–10 years) each focused on a different cause. Therefore, the school project encompassed mini-projects targeting different aspects of the problem.

How can we:

- bird proof the lunch box tubs?
- make efficient rubbish bins?
- make better lunch tubs?
- stop rubbish from going out into the community?
- stop rubbish staying on the ground?
- reduce the amount of rubbish coming into the school?

Acting on their concerns about protecting the environment, class groups conducted specific research to devise targeted solutions to minimise litter at the school. They explored materials and design solutions to reduce waste and built prototypes to assess their effectiveness in waste reduction. The student-centred inquiry gave them freedom to take control of their own learning and develop design thinking skills. For example, one class explored how to keep birds out of unattended lunch boxes in the playground. Using a range of mathematics, science and technology knowledge and skills students considered size, shape, volume, internal compartments, and bird-proof fasteners and auto-closures. "My prototype, it has a button [on the flat lid] and when you press, it slides sideways, and when you press it again it slides back, so no rubbish can get out and no birds can get it" (Student 1). Others incorporated Velcro fasteners that would be hard for the birds to open, and one prototype had an electric bird alarm activated by the lid opening. Research informed their designs "…according to our data I've seen that the sparkles do not attract them because it shines off the light. I have also researched that the white colour - they try to avoid it" (Student 2).

As a culminating activity, the students made a video to communicate their project, with each group explaining and justifying the processes and solutions represented by their prototypes.

9.5 Conclusion and Recommendations

The story of the STEM Teacher Enrichment Academy Primary program, like the STEM story of each our participating schools, is unique. However, we suggest there are some aspects of our approach to STEM education that may be transferrable to other professional learning initiatives. In essence, the STEM Academy models to the participating teachers the same approach we are encouraging them to use with their students. We seek to empower the teachers to make contextualised, creative decisions about STEM education in their schools by supporting the development of their knowledge and skills over time. The teachers are challenged by the aspirational, open-ended nature of the Academy's goals, as communicated through the two continua of integration and pedagogy and the six characteristics of STEM projects, but these high expectations are balanced with the scaffolding provided throughout the professional learning experiences. We believe that the tri-level approach for building STEM skills amongst students also serves to build the skills and confidence of the teachers. This stepped approach facilitates progress towards an integrated STEM project while allowing teachers time to make connections with their existing effective practices, draw on the varied expertise of their team members and negotiate the constraints of their school's context. The STEM Academy team strongly believes that growing a sustainable culture of STEM education in a school cannot be achieved by supplying generic pre-prepared units of work to teachers. Instead, we advocate supporting the professional growth of teachers and respecting their capacity to nurture the growth of their students. Therefore, it seems appropriate to end our story with words of a team of teachers who participated in the STEM Academy:

> Our STEM Academy story has been a lived experience for teachers and students … A good story has a message, and our message is that STEM has the potential to engage creative desire and teach above and beyond the horizon of the syllabus. The real beauty of the story was in the process. Each element of the design thinking process brought purpose, connection, authenticity, ideas and growth. For us success came from having a great team, genuine relationship with community and delving deep into the design thinking cycle ... STEM needs an authentic connection with community. It needs to have a result that will ripple beyond the school grounds … For us, the deep bond that developed out of this relationship was the drive the kids needed to work towards solving a problem and being engaged in the design process. The problem identified in STEM needs to be born from desire, whatever the topic, it provides the agency needed for those involved. (Teachers from Western School)

References

Anderson, J., Wilson, K., Tully, D., & Way, J. (2019). "Can we build the wind powered car again?" Students' and teachers' responses to a new Integrated STEM curriculum. *Journal of Research in STEM Education, 5*(1), 20–39. https://j-stem.net/index.php/jstem/article/download/61/47

Becker, K., & Park, K. (2011). Effects of integrative approaches among science, technology, engineering, and mathematics (STEM) subjects on students' learning: A preliminary meta-analysis. *Journal of STEM Education: Innovations and Research, 12*(5/6), 23–37.

Corrigan, D. (2020). *Implementing an integrated STEM education in schools: Five key questions answered*. Monash Education Futures. https://apo.org.au/node/307630

Dare, E. A., Keratithamkul, K., Hiwatig, B. M., & Li, F. (2021). Beyond content: The role of STEM disciplines, real-world problems, 21st century skills, and STEM careers within science teachers' conceptions of integrated STEM education. *Education Sciences, 11*, 737. https://doi.org/10.3390/educsci11110737

Desimone, L. M. (2009). Improving impact studies of teachers' professional development: Towards better conceptualisations and measures. *Educational Researcher, 38*(3), 181–199.

English, L. D. (2016). STEM education K-12: Perspectives on integration. *International Journal of STEM Education, 3*(3), 1–8.

English, L., & King, D. (2015). STEM learning through engineering design: Fourth-grade students' investigations in aerospace. *International Journal of STEM Education, 2*(1), Article 14. https://doi.org/10.1186/s40594-015-0027-7

Estapa, A. T., & Tank, T. M. (2017). Supporting integrated STEM in the elementary classroom: A professional development approach centered on an engineering design challenge. *International Journal of STEM Education, 4*(1), 1–6.

Goos, M., Grotowski, J., & Bennison, A. (2017). Inspiring mathematics and science in teacher education (IMSITE). Final report. *Enhancing the Training of Mathematics and Science Teachers*. https://research.usc.edu.au/esploro/outputs/report/Inspiring-mathematics-and-science-in-teacher/99611907202621/filesAndLinks?index=0

Goos, M., & Bennison, A. (2018). Boundary crossing and brokering between disciplines in pre-service mathematics teacher education. *Mathematics Education Research Journal, 30*, 255–275.

Hasso Plattner Institute of Design at Stanford University. (2018). *Stanford an introduction to design thinking: Process guide*. Retrieved from https://web.stanford.edu/~mshanks/MichaelShanks/files/509554.pdf

Hathcock, S., Dickerson, D., Eckhoff, A., & Katsioloudis, P. (2015). Scaffolding for creative product possibilities in a design-based STEM activity. *Research in Science Education, 45*(5), 727–748.

Holmes, K., Gore, J., Smith, M., & Lloyd, A. (2018). An integrated analysis of school students' aspirations for STEM careers: Which student and school factors are most predictive? *International Journal of Science and Mathematics Education, 16*(4), 655–675.

Honey, M., Pearson, G., & Schweingruber, H. (2014). *STEM integration in K-12 education: Status, prospects and an agenda for research*. National Academies Press.

Hudson, P., English, L., Dawes, L., King, D., & Baker, S. (2015). Exploring links between pedagogical knowledge practices and student outcomes in STEM education for primary schools. *Australian Journal of Teacher Education, 40*(6), 134–151.

Lowrie, T., Downes, N., & Leonard, S. (2017). *STEM education for all young Australians: A bright spots learning hub foundation paper, for SVA, in partnership with Samsung*. University of Canberra STEM Education Research Centre.

Maconachie, D. & McInnis, C. (2017). *Enhancing the training of mathematics and science teachers: Evaluation. Final report*. Australian Government Department of Education and Training. https://ltr.edu.au/resources/MS13-3293_MaconachieMcInnis_Report_2018.pdf

National Council. (2015). *National STEM school education strategy, 2016–2026*. Education Council.

Office of the Chief Scientist. (2014). *Science, technology, engineering and mathematics: Australia's future*. Commonwealth of Australia. Retrieved from http://www.chiefscientist.gov.au/wpcontent/uploads/STEM_AustraliasFuture_Sept2014_Web.pdf

Preston, C. (2020). Engineering science teacher creativity! STEM in action with preservice teachers. *Science and Children*, 69–75.

Preston, C., Davidson, T., Killick, S., Stephens, L., Meikle, S., & Fatkin, A. (2019). Teaching and learning STEM in context at Belrose Public School. *Science Education News, 68*(4), 29–37.

Prinsley, R., & Johnston, E. (2015). *Transforming STEM teaching in Australian primary schools: Everybody's business*. Commonwealth of Australia.

Rosicka, C. (2016). *Translating STEM education into practice.* Australian Council for Educational Research.
Sanders, M. (2012). Integrative stem education as best practice. In H. Middleton (Ed.), *Explorations of best practice in technology, design, and engineering education* (Vol. 2, pp.103–117). Griffith Institute for Educational Research.
The Stephanie Alexander Kitchen Garden Foundation. (2014). *Stephanie Alexander kitchen garden programme.* Retrieved from https://www.kitchengardenfoundation.org.au/
Treacy, P., & O'Donoghue, J. (2014). Authentic integration: A model for integrating mathematics and science in the classroom. *International Journal of Mathematical Education in Science and Technology, 45*(5), 703–718.
Tully, D. (2020). *Evaluation of the impact of the STEM Teacher Enrichment Academy 2018 and 2019 Primary and Secondary schools teacher analysis.* Unpublished report to Academic Committee, University of Sydney.
Vasquez, J. (2015). Beyond the acronym. *Educational Leadership, 72*(4), 10–15.
Way, J., Preston, C., & Cartwright, K. (2022). STEM 1, 2, 3: Levelling up in primary schools. *Education Sciences, 12*(11), 827. https://doi.org/10.3390/educsci12110827

Chapter 10
Success Stories (and Some Lessons) Around Professional Development for Primary Science Teachers

Simon Crook and Helen Georgiou

Abstract A picture of effective professional development for science teachers is emerging, with a number of educational jurisdictions now drawing on and requiring that professional development is designed in ways that are informed by evidence. However, professional development occurs in vastly different educational ecosystems, with outcomes that are varied and sometimes difficult to measure, especially for disciplinary studies at the primary school level. In this chapter, the first author (SC) draws on their experiences working in primary schools providing professional development in STEM to primary teachers, with the second author (HG) providing critical reflections. This chapter will be centred on effective professional development principles, enriched with case studies that more fully describe the evolving nature of professional development dictated by a range of factors, including teacher, student and systemic needs.

Keywords Professional development · Primary science education · STEM

10.1 Introduction

Teacher professional development (and more broadly, teacher education) is crucial to maintaining teacher well-being, adaptiveness, professionalism and ensuring a high-quality educational experience for students (OECD, 2019). Whilst what is

S. Crook
The University of Sydney, Sydney, Australia

CrookED Science, Dulwich Hill, Australia

S. Crook
e-mail: simon@crookedscience.com

H. Georgiou (✉)
University of Wollongong, Wollongong, NSW, Australia
e-mail: helengeo@uow.edu.au

S. Earle et al. (eds.), *Primary Science Learning for Children, Teachers, and Communities*, https://doi.org/10.1007/978-981-95-0561-6_10

meant by professional development might sometimes be difficult to comprehensively capture (Evans, 2002), the term can broadly be defined as '*...activities that develop an individual's skills, knowledge, expertise and other characteristics as a teacher*' (OECD, 2009, p. 49). Teacher professional development encompasses a wide range of practices, both formal and incidental, such as professional collaboration, research, mentoring, observation and knowledge sharing, that might develop knowledge, skills, identity or attitudes (Fraser, 2010). In responses to widely distributed surveys, teachers reported attending experiences such as courses or workshops, education conferences, qualifications, observation visits, engagement with professional networks or teacher associations, engaging in research and mentoring (OECD, 2009), and these could consist of one-off, short interactions or multi-year programs (Trygstad, 2013). Fraser also identifies *professional learning* as a process of becoming a more expert teacher, encompassing a wider range of activities and characteristics of development (2010). Despite this distinction, the terms 'professional learning' and 'professional development' are often used interchangeably. In this Chapter, we refer to 'professional development' when discussing organised programs targeted at teachers, but we are inclusive of the broader definition captured by the term 'professional learning' and highlight where this is relevant.

Despite the diversity and complexity of professional development experiences, a robust framework of 'best practice' has emerged from the research literature and is being utilised in practice and policy (e.g., AITSL, 2017; Darling-Hammond et al., 2017; NESA, 2022; OECD, 2019; Sims et al., 2021; Wilson, 2013). Whilst the language slightly differs between these reports, the same themes can be identified, including that high quality, impactful professional development tends to be (i) focused on specific content, (ii) of sufficient duration, (iii) exhibiting strong coherence with school policy and practice, (iv) supporting active collaboration between teachers and administrators, and (v) engaging teachers in active learning.

However, impact is difficult to measure. Part of the challenge of understanding the impact of professional development programs is the huge variety of programs experienced by teachers (Sims et al., 2021). These programs are different lengths, focused on different discipline areas, educational stages, topics and use a variety of different approaches. In Australia, professional development in the key learning area of science at the primary level is not common (Deehan & MacDonald, 2024). Deehan and MacDonald (2024) report that amongst their sample of 206 primary teachers, the majority (62%) reported not receiving *any* science professional development: "This is of concern as the educators … had been teaching for over 15 years on average" (p. 92). A rich description of professional development experienced in primary schools is therefore sorely needed. It is through this lens that the first author of this paper (SC) provides the following reflections. These reflections are synthesised, critically examined and situated with the literature by the second author (HG). The aim of this chapter is therefore to provide detail to these effective professional development practices in an area that otherwise is not well described in the context of primary science education, highlighting the possibilities within this space.

10.2 Background

Primary science, like the many other key learning areas in the primary space, is characterised by distinct pedagogies, learning environments, triumphs, and challenges (as described in this collection). It is widely acknowledged that primary teachers report feeling unprepared to teach science at the primary level (Banilower et al., 2018; Trygstad, 2013), often understood as a consequence of the focus on the 'core' learning areas of mathematics and literacy. For instance, in a survey distributed amongst United States (US) elementary teachers, 73 and 77% of teachers felt well prepared to teach mathematics and literacy, respectively, whilst only 31% felt prepared to teach science (Banilower et al., 2018). The proportions become more extreme for specific subjects, such as physics and engineering, which are often taught together in primary science, or 'STEM' (Banilower et al., 2018). Low self-efficacy or confidence towards science are known to lead to less attention being paid to it in the classroom, resulting in poorer outcomes for students (Haverly & Davis, 2024). In addition to these challenges, science is a practical subject, with investigation skills being at the centre of teaching and learning.

For these reasons, professional development in primary science is vital. In science education, professional development also has some unique characteristics. For instance, results from science-specific professional development at the primary level reveal that primary science teachers most frequently engaged in activities such as participating in science investigations, collaborating with colleagues in the same school to plan lessons, experimenting with new teaching approaches in the classroom, and reflecting on classroom artefacts (Trygstad, 2013). Surahman and Wang (2023) report on a science-specific systematic review revealing that constructivist theory was the most widely adopted learning theory, with the most common professional development being the workshop, training and online models. They highlight a particular need to involve school administration and leadership. Thus, in addition to the principles of effective professional development outlined above, effective professional development in science also includes immersing teachers in inquiry experiences and witnessing models of inquiry teaching (Wilson, 2013).

Haverly and Davis (2024) describe primary science teachers as exhibiting assets such as "inquisitiveness, an orientation towards inquiry, understandings of inquiry, capacities for learning through teaching, and knowledge of instructional approaches" (p. 106), which can be leveraged by both teacher educators and professional development programs. A strengths-based approach which acknowledges primary teachers' diversity and their commitment to students and their practice is therefore vital in order to support high quality teaching practices and highlight the potential of positive professional development experiences.

10.3 Method

The first author (SC) is a former high school science teacher (of 15 years), postgraduate researcher, and professional development provider (of 17 years). It is with the lived experience of delivering professional development focused on primary science, and the experience SC gained as a researcher, that he, together with HG, present this case study, which is centred on personal reflections (SC), and enriched by critical engagement with the extant literature (HG) to connect personal experience to wider meanings and understandings, in this case, in the area of professional development in primary science.

The reflective, autoethnographic-inspired process involved writing about SC's personal experiences of professional development in primary science over the course of recent years (2021–2024). These writings resemble journal entries or narratives which explore observations that seemed noteworthy during the process of delivering professional development in the primary school sector. The focus, generally, was to capture the essence of professional development, but with a strengths-based focus; to explore what worked and why. In a systematic and cumulative process, SC shared these observations with HG, and together, we processed these understandings, incorporating knowledge from the extant literature and engaging in a reflexive process of questioning, critically examining and reframing these experiences for a wider audience (Adams et al., 2015).

Ultimately, we decided to organise the reflections under four of the six key principles of effective professional development in primary science, drawn from the literature summarised above, and these reflections form part of the 'results and discussions' section of this chapter. The four specific principles chosen represented those identified as most relevant to primary science professional development in the literature. The rationale for this was the increasing attention paid to 'evidence-based' practice in Australia and the need to situate SC's reflections within this literature.

10.3.1 Ethical Perspectives

Due to the self-reflective nature of this research, the work was not approved by an ethics board (nor did we seek approval). However, we took considerable efforts to consider any ethical implications of the work in accordance with the Australian 'National Statement' (NHMRC, 2023). To address any real or perceived conflict of interest, SC is transparent about the fact he is a business owner of a professional development organisation. However critically SC might engage in the reflective process, there is always a risk of some bias in these reflections. For this reason, the role of the second author, HG is important, as it offers balance and criticality.

Another ethical quandary is whether this research involved participants, and to what degree these participants have been denied the ability to provide consent to their 'data' being used. Our solution to this issue is to, as much as possible, focus on

SC's experiences and observations. To avoid issues of identification, some contextual details have been omitted, and when specific examples are mentioned, details will be blended from more than one individual or school together, to form the reflection. Some materials in this chapter (e.g., images, quotations) are also provided, for which express permission to freely use was received.

10.3.2 Context

This chapter is focused on exploring professional development success in primary science. To fully understand the phenomenon explored, some detail about the context of this work is required. The work is undertaken in the state of New South Wales (NSW) in Australia, where, at the time of writing this chapter, teachers must complete 100 hours of professional development in a five-year period to maintain their teacher registration, with 50 hours of this allocation needing to be specifically accredited by the NSW Education Standards Authority (NESA). This requires professional development providers to submit applications that respond to various criteria and are based on the principles discussed above.

Most of the schools SC works with belong to the Independent school sector. In Australia, around 65% of students attend Government (public) schools, with the remainder in Independent schools. Because Independent schools are governed by smaller and more regional authorities (rather than the single authority that governs public schools for each state), they are (theoretically) more flexible in terms of engaging professional development providers and educational researchers.

The main form of professional development SC provides is the 'primary partnership' (PP) for individual primary schools. This involves engaging with staff once a fortnight to co-develop units and team- or co-teach science lessons. In addition to the primary partnerships (PP), online on-demand courses and professional development workshops are offered.

Australia has a national curriculum, called the 'Australian Curriculum' (AC). Whilst all states and territories in Australia are required to adhere to the AC, each individual state has the authority to wholly 'adopt', or 'adapt' it. NSW is an 'adapter' state, meaning they have a state syllabus which is based on the AC, but not identical to it. The current primary science curriculum, based on the AC, was published in 2017 and implementation began from 2018 (NESA, 2017). However, this replaced a previous syllabus which was only recently introduced in 2015.

The current syllabus, known as the 'NESA K-6 Science and Technology syllabus' or just 'NSW syllabus', is divided into four stages: Early Stage 1 (Kindergarten, age 5/6), Stage 1 (Grades 1 and 2, age 6–8), Stage 2 (Grades 3 and 4, age 8–10) and Stage 3 (Grades 5 and 6, age 10–12). The syllabus includes three different key learning areas from the AC, including Science, Digital Technologies and Design and Technology. This means that science teaching at the primary level in NSW is inherently interdisciplinary. In particular, the 'Design and Production' skills are expected to be integrated with science content (e.g., making shelter for a living thing, making devices with electric circuits). As such, the terms 'science' and 'STEM' are used interchangeably in this chapter. The syllabus is based on outcome statements that allow for considerable flexibility, though recent curriculum reform in Australia has resulted in reduced flexibility, with the new syllabus (due for publication in 2024) consisting of more prescriptive outcome statements and associated support material. There are no compulsory national science tests in Australia for all primary students (Georgiou, 2023).

10.4 Results and Discussion

The following section is divided into different characteristics of effective professional development in primary science education, including focusing on (i) specific content, (ii) professional development that is of sufficient duration, (iii) facilitating collaboration and including (iv) immersion in inquiry experiences. These sections are enriched with reflections from the first author's experiences providing professional development and classroom support across a wide range of different primary schools, and supported by the second author's reflections and references to the extant literature.

10.4.1 Focusing on Specific Content

Professional development is most effective when it focuses on both content and pedagogy associated with curriculum content. Ideally, this professional development should be strongly related to teachers' classroom practice and the students they teach, rather than generic professional development which is independent of context (Darling-Hammond et al., 2017).

Because detailed research on professional development for primary science is rare, there is a wide range of possibilities when considering 'focusing on specific content'. In this section, we outline three key activities: leveraging existing knowledge and expertise whilst providing 'in the moment' support; teaching content in authentic contexts; and, working with teachers to coherently adjust and integrate pre-packaged materials into their school curriculum.

Primary teachers of Science are known to exhibit a lack of confidence when interpreting curriculum documents (Schwarz et al., 2008). A key opportunity in the professional development SC provided, related to this, was to support teachers through a period of curriculum reform by leveraging and capitalising on their existing planning materials. In NSW, teachers experience frequent syllabus changes, and there was a tendency by some teachers to either start afresh. For example, in the 2015 syllabus, one Stage 3 'Physics' outcome related to exploring electric circuits, whilst the corresponding 2018 syllabus outcome foregrounded energy transformations in circuits (NESA, 2017). Although the tendency was to discard previous units to create

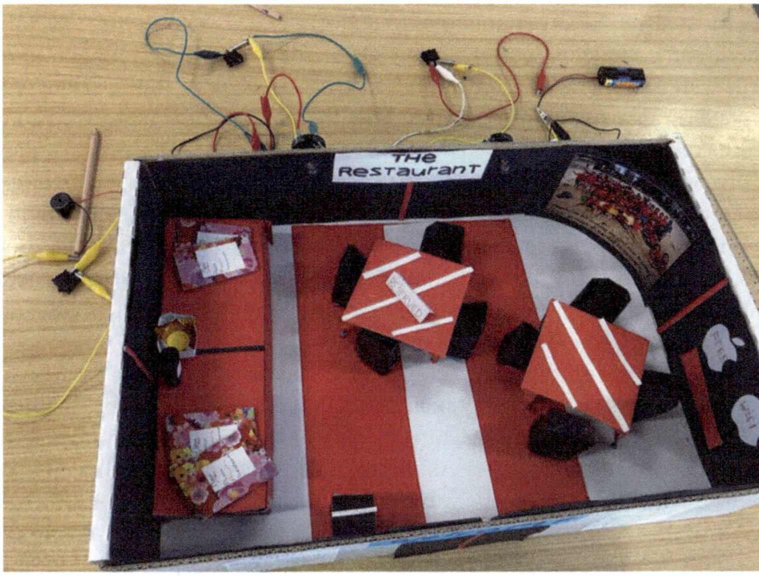

Fig. 10.1 A model restaurant created by Grade 6 students involving an electric circuit for a 3D space

new ones, as part of the PP program, SC reflects on working with teachers to instead to adapt the learning units to maintain the deep exploration of electric circuits (how they work, the different components, how to make a circuit, how to draw a circuit etc.), but with more explicit emphasis on energy transformations. This 'in-the-moment' support is consistent with reports from the literature, that highlight the need to provide ongoing support to primary teachers of science during a period of curriculum reform (Schwarz et al., 2008). Figure 10.1 shows a model restaurant with a parallel circuit, each parallel with its own switch, for lights, a buzzer (modelling the audio system) and a motor for the ceiling fan.

Another key example of focusing on specific content, was the work with teachers to embed content in local, authentic contexts. Teaching in context has been advocated in primary science education for decades, as a way to engage students and teach science authentically (Diamond et al., 2014; Rodrigues, 2006). However, authentically connecting real-world contexts to science content and skills is not straightforward, and if students are not involved in this process, the connections may not be successful (Rodrigues, 2006). One of the strengths of this current syllabus was its flexibility, and that there were ample opportunities to authentically adapt to local contexts and maintain relevance to local teachers and students. As part of the PP program, a collaboration was established with teachers *and* students to embed authentic contexts within units of work. For instance, one school was undergoing a significant building program which displaced Stage 2 classes into demountable (temporary) classrooms on the school field. This field had quite a steep slope, and after two seasons of unusually persistent rain, the field was transformed into a quagmire, requiring students and teachers to wear gumboots (rainboots) to and from their classrooms. The following year, these same students were learning about soil, erosion and human impact on the Earth's surface. Drawing on the muddy experiences of the previous year, we co-designed an investigation to explore how the amount of water absorbed by soil changed its consistency (and hence had the potential to make for a very muddy existence). Five tubs were laid out filled with the same amount of soil. Different measurements of water were added to different tubs. Next, nominated students would put on some gumboots and stomp around in each tub, describing and scoring the experience, with the soles of the boots examined each time. The students loved this activity—everything from working outside, to conducting a hands-on fair test with soil and water, to stomping around in mud in the name of science.

A second example of focussing on specific content within local contexts occurred with a Stage 3 class in two very small schools that were due to be amalgamated. As part of the amalgamation, one of the schools underwent a significant building program to become the new larger combined school. The school principal expressed a desire to incorporate a 'sensory garden' in new facilities (Fig. 10.2). At the time, Stage 3 in both schools were exploring the functional and structural properties of different materials in the local environment. Accordingly, as part of the PP program, we embarked on a collaborative project between the two small schools engaging both Design and Production, investigation skills (NESA, 2017) and science content knowledge. This project followed the design process, from presenting the challenge of creating a sensory garden for the new amalgamated school, researching the shape and size of the plot for the sensory garden through discussion with the principal and study of the blueprints, liaising with an architect, creating initial designs, making 3D models and presenting their finalised designs to each other. The architects responsible for the project then incorporated design elements from the winning projects into the final product. Students and teachers were invested and took ownership of the sensory garden, which remained a source of pride and connection within the amalgamated school.

As Rodrigues (2006) explains: "If the contexts are to be appropriate whereby we expect students to explore and relive their everyday world … then we should expect teachers to explore students' thinking and we should provide real opportunities for pupils to explore scientists' thinking" (pp. 181–182).

Final reflections related to 'specific content' involve supporting teachers to meaningfully incorporate pre-packaged curriculum programs into their school-based units of work. In Australia, these include the freely available *Primary Connections*[1] or paid resource banks of such as *Inquisitive*[2] and *TeachStarter*,[3] and it is common for most teachers to adapt existing lesson plans and units rather than create wholly new ones. *Primary Connections* is an initiative of the Australian Academy of Sciences, created and sustained from considerable funding from the federal Government (approximately $15 million). At the core of the program sits several all-inclusive curriculum

[1] https://primaryconnections.org.au/.

[2] https://www.inquisitive.com/au.

[3] https://www.teachstarter.com/au/.

Fig. 10.2 Grade 5 students from two schools in a video conference with architects regarding the design of a sensory garden

units, though these are intended to be implemented through participation in a professional development program (Hackling et al., 2007). Whilst Primary Connections was a huge success and had significant reach, there were challenges that only intensified in NSW due to the significant curriculum reform in that state. This is because Primary Connections was based on the AC, whilst NSW was engaging in curriculum reform which meant that its syllabuses looked considerably different than the AC (additional topics and different content focus). Combined with an overall shift to digital resources (AC was at the time predominantly paper or PDF based), and the use of these resources in isolation (without the professional development), implementation became less effective and teachers' noted frustrations (Aubusson et al., 2019). Anecdotally, teacher expertise was not being developed as expected when using externally produced units of work, particularly in terms of the scientific inquiry skills required to design and run investigations. As recognised by the Primary Connections professional development, teachers require some contextual understanding and professional development support to deliver pre-packaged resources effectively (Aubusson et al., 2019). However, the view when using or purchasing these packages is that they are 'plug and play'. As part of the PP, engaging with teachers to 'cherry pick' resources was undertaken, in an effort to develop more suitable tailor-made resources.

Teachers felt much more empowered understanding the purpose of the units (e.g., the literacy focus and the '5Es') and could more readily adapt individual resources across teaching contexts with this support e.g. adapting to the NSW scientific inquiry skills known as 'Working Scientifically'. (It should be noted that since the time of these issues, Primary Connections has had a massive overhaul, with everything moving online, and without the prerequisite PD. Interestingly, SC pointed out some minor errors in Primary Connections to them, which they fixed in a very efficient and satisfactory manner).

10.4.2 Sufficient, Lasting Duration

Another key principle for effective professional development is that it must be of sufficient and lasting duration. Although the threshold for this has not been established, research has shown that one-off workshops and short professional development programs are unlikely to deliver any immediate or long-term results. Experiences should instead be sustained and involve multiple opportunities for engagement around individual themes (Darling-Hammond et al., 2017).

Working with teachers over longer periods of time (one to five years), the challenges posed by ultra-short course offerings, and the benefits of sustained professional development are clear (SC). One of the key issues in primary science investment was not necessarily the engagement in one-off courses (as there aren't very many offered), but more one-off investments in equipment without investing in developing teacher expertise. Another observation in relation to primary science professional development was that it was difficult for teachers and school leaders to articulate their needs. Further, these needs changed significantly over time, recognising the need for the professional development to be adaptive. These reflections will be outlined further below.

One of the most impactful observations that occurred while visiting NSW primary schools was the significant investments in expensive STEM resources. These resources, being marketed at teacher conferences or seen at nearby competitor schools, were often related to 'digital technologies' (e.g., robots) and to satisfy the requirements of the new syllabus (NESA, 2017), which included digital technologies content (e.g., coding, data) for the first time. Too often, these expensive resources would arrive and be rarely used, if at all, even after the associated expensive one-off professional development. When key early adopter colleagues subsequently moved on to different schools, these resources usually were not used at all by the remaining teachers and simply occupied space on shelves and in cupboards. This is consistent with research which highlights the barriers to integrating educational technologies such as robotics. In particular, teachers report needing significant support, both technical and pedagogical to incorporate robotics into their classroom teaching, despite being interested or enthused to do so, and integration remains low (Khanlari, 2016; Negrini, 2020).

A far more effective and versatile approach is to encourage purchasing everyday items cheaply, from a supermarket (e.g. bicarbonate soda, ziplock bags, spaghetti, food colouring and more). Should more technical science resources be required, it would be helpful to primary schools to be provided with materials that were both competitively priced from reliable suppliers and appropriate to primary school needs (e.g. not glass yet durable). If a school had invested in new technologies that weren't being used, such as 3D printers, Spheros or Bee-Bots, the recommendation would be that they didn't purchase anything further until these were in use and integrated in the teaching programs. Often, how to use these technologies in STEM became a focus of the professional development offered over the year(s). Ongoing support helps develop teacher confidence, and support during the first implementation is often enough for teachers to be able to continue use (Khanlari, 2016; Negrini, 2020). For example, in one case, where Kindergarten teachers were teaching forces and motion in one unit and Bee-Bots in a separate Digital Technologies unit, they were encouraged to combine the two and teach forces and motion *through* Bee-Bots. This led to a far more authentic integration and application of Bee-Bots which teachers (and students) appreciated. Adding the app to iPads also helped alleviate any issues created with equipment shortage or sharing.

A second, and extraordinary observation that occurred, was that, beyond adapting to the new syllabus, schools and teachers' need for science professional development was not well articulated. It was clear at the outset that schools didn't really know what they wanted or required beyond making science and STEM 'alive' in their school. Given the professional development work offered involved sustained collaboration, the challenge was to ensure each experience was adaptive and relevant to each individual school context. The PP program, which involves a minimum one-year commitment, and commonly extends beyond this, is unique for the primary sector. This work involved providing teachers and schools with the language to articulate and have their needs met in the professional development space. Its structure facilitates a 'professional learning' model, where teachers have the opportunity to engage in professional development 'activities', but also reflect on their own learning as part of a professional community (within the school), between these more formal activities. The model of sustained interaction is consistent with a view of professional learning as encompassing 'formal, informal, planned, and incidental opportunities' (Fraser, 2010, p. 93).

The PP model arose out of discussions with school principals who recognised the need to embed practices over time. Initially, this involved establishing relationships in schools with teachers and students by delivering lessons in K-6, to make science fun and engaging, and demonstrating best practice (Wilson, 2013). Co-planning and co-teaching were generally considered to be too overwhelming in these initial stages. Instead, the practices of mentoring and observing provided teachers with the confidence that they could do it themselves, and then become school leaders in primary science. After a period of demonstration, the PP program developed into a model which allowed more effective capacity building. The resulting program took the form of three distinct stages over another year: co-planning with individual teachers to look at the content of their next unit and their existing teaching programs, then co-designing an engaging lesson to team teach, then co-designing an engaging lesson to team teach with the teacher leading. This was achieved by teachers being released from their teaching to have meeting time, with the collaborative co-development of units achieved through Google Docs (and more recently Google Slides). Interestingly, the research on professional development, particularly in relation to longer engagements and sustainability, primarily focuses on measurements of key outcomes after the program has ended. Beliefs are considered more difficult to change than practices, and in most cases, these practices and beliefs are pre-determined (and therefore measured) (e.g., Drits-Esser et al., 2017; Sandholtz & Ringstaff, 2016; Wolf & Peele, 2019). There is much less exemplification of a collaborative approach between teachers and professional development providers, resulting in a lack of understanding of the changes that might occur over longer periods of engagement.

The program was received (and continues to be received) very positively, with the long-term benefits being particularly highlighted by school leaders and teachers:

> The sessions combined what we were already doing in the classroom as well as what we could change to meet the needs of all students. They enabled us to develop valuable assessments and create differentiated and meaningful programs. (teacher)
>
> This professional learning helped increase engagement with all students and the evidence of language used by the students. (school leader)

10.4.3 Collective Participation

Collective participation, or collaboration between different school-based stakeholders, has been identified as an important quality of well-designed professional development. Collaboration might involve small group or one-on-one interactions, to schoolwide interactions, to collaborations involving a variety of configurations spanning whole districts and external stakeholders (Darling-Hammond et al., 2017).

In this section, the benefits of collaborative and collective participation in professional development for primary science teachers are outlined. Different kinds of collaboration are exemplified here, including working collectively with colleagues, and engaging administrators and school leaders.

In the first instance, teachers engaging in the PP program have consistently been incredibly positive about opportunities for collaboration. Often, teachers would be thankful simply for the opportunity to work and talk together, as their busy schedules often meant that there was little time for this, echoing the report from Goodrum (2001 as cited in Jones, 2008), that 'Teachers frequently report that they lack the time and opportunity to share ideas, collaborate, reflect, evaluate, adequately prepare and participate in ongoing learning/professional development;' (p. 87). In the primary sector in NSW, primary teachers are not away from their classes for long periods, receiving only two hours of 'relief' from face-to-face teaching each week. In a regular week, although this time is intended for longer term planning, the urgency of many day-to-day activities will often consume this allocation, leaving the teachers without any opportunity to properly reflect on their practice. Teachers appreciated the way in which collaboration was facilitated, often when working with the same colleagues, an external guide is more effective than self-management.

However, these opportunities to engage teachers in collective collaboration would not be possible without school administrators and leaders' support. Engaging school administrators and leaders is fundamentally important because without their cooperation and buy-in, any professional development initiative can be undermined (Lipscombe et al., 2020). For instance, teachers, feeling overwhelmed with day-to-day responsibilities might be tempted to skip or disengage from external professional development during their relief allocation. Further, there are schools with competing initiatives such as sport, music, gifted mathematics programs etc., which often involve other external providers and the associated investment and accountability. Good school administration is required to organise and enforce an overall school timetable such that clashes do not occur, and teachers and students aren't left having to choose between one activity and another. Administrators require value for money and meeting key performance indicators (KPIs) from all potentially competing initiatives.

System level administrators benefit from being informed of and made sensitive to local professional development initiatives. For example, one school was involved in a system-wide external initiative around integrating and using 3D printers in learning. This required timetabling organisation to facilitate, and the cost of covering the teachers involved. Some of the same teachers had clashes between this and the PP program. Accordingly, we were able to capitalise on this situation thanks to good administration by firstly adapting the PP program to include integrating 3D printing in science, then collaborating asynchronously due to the timing clashes, such that remote support was provided during working hours, and the teachers caught up after hours (recompensed with time in lieu). Strong, creative leadership empowering flexible professionals allows for much greater possibilities in schools which might otherwise be hamstrung by convention.

Perhaps most importantly, school leaders need to be made aware of the curriculum and practical needs of science and STEM. Science professional development for primary school principals and leaders is one example of how to facilitate this, and SC offered workshops which engaged these leaders skills' based activities. Leaders often comment afterwards on how their everyday work distances them from the curriculum and the classroom. They expressed how much they enjoyed getting hands-on experience and seeing what their teachers and students go through. In addition, the school leaders gain a greater appreciation of the limitations of packaged resources and the potential detriment they may have on teaching and learning if adopted as the be all and end all.

As outlined by Lipscombe et al. (2020), collaboration in schools is part of an open system involving inputs, throughputs, and outcomes. This complex system implicates the role of administration, policies, and school leadership, amongst others, as an individual or team's actions are impacted by, and in turn, impact the organisation. For professional development to be successful, a cohesive, shared vision and deep interactions are considered necessary.

10.4.4 Immersed in Inquiry Experiences

Inquiry is considered a central pedagogy or philosophical approach to teaching science. Although the term is contested, it relates to the process of 'doing' science and is usually associated strongly with undertaking scientific investigations. Inquiry practices feature heavily in science professional development and have been associated with more positive professional development outcomes (Wilson, 2013).

In this case, several examples are provided of engaging science teachers in the scientific inquiry process and learning how to foster student scientific inquiry. These range from whole school staff development days to individual collaborative planning.

Inquiry-based science education is taught extensively in initial teacher education courses, and exhibits benefits for students, but is often challenging to implement (Strat et al., 2024). In the context of this work, the definition of inquiry used is consistent with that of the National curriculum authority, the Australian Curriculum and Assessment and Reporting Authority (ACARA, n.d.), which focuses on inquiry as the process involved in undertaking scientific investigations (rather than Inquiry as a pedagogy). ACARA emphasises that 'the skills students develop give them the tools they need to achieve deeper understanding of the science concepts and how scientific thinking applies to these understandings' and that inquiry skills should be integrated with other content. This supports an approach for professional development that is based on developing skills alongside with content, leveraging the skills involved in investigations, which are often more accessible to teachers and students, to develop conceptual understanding (Strat et al., 2024). For example, with the implementation of the then NSW syllabus in 2018, a focus on the scientific inquiry skills could act as an anchor across the old and the new syllabuses. A key initiative developed as part of the professional development offering, based on this understanding, was the *Engaging K-6 Science & Technology Staff Development Day,* which was hands-on, engaging and fun, while ensuring that all staff learned the fundamental principles of running a 'fair test' and how to teach this. The day also emphasised the importance of skills over content at the primary level. An aspect that shocked but also gratified teachers was the provision of a blueprint for integration of skills and content, allowing them to find more 'space' in an otherwise crowded curriculum. Interestingly, (but anecdotally) many teachers would tend to overburden their programs with unnecessary and inappropriate content knowledge (such as Newton's Laws in Grade 2). As a consequence of teachers learning the scientific inquiry skills, they were more comfortable with running and even designing hands-on fair tests. Students would report back about the fun, engaging experiments they had conducted with their teachers, and the science they therefore had learned.

At the other extreme, more recent work has involved withdrawing gifted students directly to work on year-round scientific inquiry projects as part of the Science Teachers' Association NSW 'Young Scientist' competition.[4] While the focus is on the students, this works best when there is also a liaison teacher from the respective school, checking in on the students in between visits and maintaining the momentum

[4] https://stansw.asn.au/YoungScientist/YoungScientist/YS-Home.aspx.

of the students' projects. The projects involve students engaging in every aspect of the scientific inquiry cycle, several times over, as they conduct a range of experiments around a central theme (e.g., investigating the effects individual variables have on projectile motion). In one such school, whilst working with in Grades 4–6, the liaison teacher follows and replicates this work Grade 3. In her words, her work with Grade 3 this year is so much better than previous years having observed and then engaged in inquiry teaching, demonstrating the impact of modelling inquiry processes (Wilson, 2013). Watters (2021) highlights the importance of inquiry projects in science for gifted students. Crucially, he explains that they offer a unique opportunity for students to create new knowledge and engage them in future STEM careers.

10.4.5 Conclusions and Implications

These stories are shared with the intent of providing detail to the otherwise broad set of principles usually used to describe high quality professional development. In the first instance, such stories should inspire confidence. Drawn from experiences in primary classrooms and directly with primary teachers, these reflections may motivate, boost self-belief, and provide reassurance amongst teachers who are known to be predisposed to accepting a deficit view of their practice. For instance, such stories can also act as a motivator for self- and peer-reflection, developing self-awareness and engaging in a culture of collaboration that can ultimately have positive impacts on both teachers and students. For instance, educators might feel reassured from understanding that a 'focus on specific content' can mean leveraging their own expertise and teaching ideas, supported by 'in the moment' advice.

Beyond impacting individual teachers, these reflections are also relevant to professional development providers. The sharing of successes and challenges helps further shape our understanding of what 'effective' professional development may involve. For example, the benefits of professional development 'of lasting duration' go beyond achieving specific outcomes; a longstanding professional development experience allows adaptiveness and tailors programs to teachers' needs, more akin to a 'professional learning' model. Many professional development programs are designed a-priori, rather than considering how the program might adapt with input from teachers as partners.

The reflections also have implications further afield. School leaders, who admit experiencing increasing distance from the classroom, might find specifics relating to the mechanics of longstanding professional development relationships useful. In particular, the notion that professional development does not occur in isolation, but that school-level factors are both affected by and themselves affect an individual teacher's professional development experience. Ambitiously, revealing the paucity of professional learning opportunities for primary science amidst these success stories might even impact policy; perhaps a minimum number of professional development hours per subject would be appropriate, or increased funding to make high quality professional development in primary science more accessible.

References

ACARA. (n.d.). *Science inquiry skills*. https://australiancurriculum.edu.au/f-10-curriculum/science/structure/

Adams, T., Jones, S. L. H., & Ellis, C. (2015). *Autoethnography: Understanding qualitative research*. Oxford University Press.

AITSL. (2017, October). Uncovering the current state of professional learning for teachers. *Summary findings report*. https://www.aitsl.edu.au/docs/default-source/hqpl/aitsl-hqpl-summary-findings-report-final.pdf

Aubusson, P., Skamp, K., Burke, P. F., Pressick-Kilborn, K., Ng, W., Palmer, T., Goodall, & A., Fergusson, J. (2019). *Primary connections: Linking science with literacy Stage 6 research evaluation final report*. Prepared for Steering Committee of Primary Connections, Australian Academy of Science. https://opus.lib.uts.edu.au/handle/10453/134515

Banilower, E. R., Smith, P. S., Malzahn, K. A., Plumley, C. L., Gordon, E. M., & Hayes, M. L. (2018). *Report of the 2018 NSSME+*. Horizon Research Inc https://eric.ed.gov/?id=ED598121

Darling-Hammond, L., Hyler, M., & Gardner, M. (2017). *Effective teacher professional development*. Learning Policy Institute.

Deehan, J., & MacDonald, A. (2024). Relations between Australian primary teachers' reported professional development engagement and their science teaching practices and efficacy beliefs. *Australian Journal of Teacher Education, 48*(8), 79–118.

Diamond, B. S., Maerten-Rivera, J., Rohrer, R. E., & Lee, O. (2014). Effectiveness of a curricular and professional development intervention at improving elementary teachers' science content knowledge and student achievement outcomes: Year 1 results. *Journal of Research in Science Teaching, 51*(5), 635–658.

Drits-Esser, D., Gess-Newsome, J., & Stark, L. A. (2017). Examining the sustainability of teacher learning following a year-long science professional development programme for inservice primary school teachers. *Professional Development in Education, 43*(3), 375–396.

Evans, L. (2002). What is teacher development? *Oxford Review of Education, 28*(1), 123–137.

Fraser, C. A. (2010). Continuing professional development and learning in primary science classrooms. *Teacher Development, 14*(1), 85–106.

Georgiou, H. (2023). Are we really falling behind? Comparing key indicators across international and local standardised tests for Australian high school science. *Research in Science Education, 53*(6), 1205–1220.

Hackling, M., Peers, S., & Prain, V. (2007). Primary connections: Reforming science teaching in Australian primary schools. *Teaching Science, 53*(3), 12–16.

Haverly, C., & Davis, E. A. (2024). Unpacking readiness for elementary science teaching: What preservice teachers bring and how that can be shaped through teacher education. *Studies in Science Education, 60*(1), 75–119.

Jones, M. M. (2008). Collaborative partnerships: A model for science teacher education and professional development. *Australian Journal of Teacher Education (Online), 33*(3), 61–76.

Khanlari, A. (2016). Teachers' perceptions of the benefits and the challenges of integrating educational robots into primary/elementary curricula. *European Journal of Engineering Education, 41*(3), 320–330.

Lipscombe, K., Buckley-Walker, K., & McNamara, P. (2020). Understanding collaborative teacher teams as open systems for professional development. *Professional Development in Education, 46*(3), 373–390.

Negrini, L. (2020). Teachers' attitudes towards educational robotics in compulsory school. *Italian Journal of Educational Technology, 28*(1), 77–90.

NESA. (2017). *K-6 science and technology syllabus*. https://educationstandards.nsw.edu.au/wps/portal/nesa/k-10/learning-areas/science

NESA. (2022). *The quality of teacher professional learning*. https://www.nsw.gov.au/sites/default/files/2022-11/Quality-of-teacher-professional-learning.pdf

NHMRC. (2023). *National statement on ethical conduct in human research.* https://www.nhmrc.gov.au/about-us/publications/national-statement-ethical-conduct-human-research-2023#block-views-block-file-attachments-content-block-1

OECD. (2009). *Chapter 3 The professional development of teachers in creating effective teaching and learning environments: First results from TALIS.* https://www.oecd-ilibrary.org/education/creating-effective-teaching-and-learning-environments_9789264068780-en

OECD. (2019). TALIS 2018 results (volume I): Teachers and school leaders as lifelong learners. *TALIS, OECD Publishing.* https://doi.org/10.1787/1d0bc92a-en

Rodrigues, S. (2006). Pupil-appropriate contexts in science lessons: The relationship between themes, purpose and dialogue. *Research in Science & Technological Education, 24*(2), 173–182.

Sandholtz, J. H., & Ringstaff, C. (2016). The influence of contextual factors on the sustainability of professional development outcomes. *Journal of Science Teacher Education, 27*, 205–226.

Schwarz, C. V., Gunckel, K. L., Smith, E. L., Covitt, B. A., Bae, M., Enfield, M., & Tsurusaki, B. K. (2008). Helping elementary preservice teachers learn to use curriculum materials for effective science teaching. *Science Education, 92*(2), 345–377.

Sims, S., Fletcher-Wood, H., O'Mara-Eves, A., Cottingham, S., Stansfield, C., Van Herwegen, J., & Anders, J. (2021). *What are the characteristics of teacher professional development that increase pupil achievement? A systematic review and meta-analysis.* Education Endowment Foundation. https://educationendowmentfoundation.org.uk/education-evidence/evidence-reviews/teacherprofessional-development-characteristics

Strat, T. T. S., Henriksen, E. K., & Jegstad, K. M. (2024). Inquiry-based science education in science teacher education: A systematic review. *Studies in Science Education, 60*(2), 191–249.

Surahman, E., & Wang, T. H. (2023). In-service STEM teachers professional development programmes: A systematic literature review 2018–2022. *Teaching and Teacher Education, 135*, Article 104326.

Trygstad, P. J. (2013). *2012 National survey of science and mathematics education: Status of elementary school science.* Horizon Research Inc.

Watters, J. (2021). Why is it so? Interest and curiosity in supporting students gifted in science. In *Handbook of giftedness and talent development in the Asia-pacific* (pp. 761–786).

Wilson, S. M. (2013). Professional development for science teachers. *Science, 340*(6130), 310–313.

Wolf, S., & Peele, M. E. (2019). Examining sustained impacts of two teacher professional development programs on professional well-being and classroom practices. *Teaching and Teacher Education, 86*, Article 102873.

Part III
Primary Science Learning for Communities

Chapter 11
Moving Beyond the Deficit Paradigms of Multigrade Teaching in Rural Primary Science Classrooms

Sphamandla Zulu, Thabisile Nkambule, Doras Sibanda, and Ange Fitzgerald

Abstract In this chapter, we examine the notion of multigrade teaching in rural primary schools with the focus on how primary science teachers manage the associated complexities. We argue that to ensure effective multigrade teaching, teachers must possess specialized skills in planning, organising instruction, and assessing both the content knowledge and content skills in the diverse, heterogeneous classrooms. In this chapter, we report on the findings of a bigger project that observed Grade 6 lessons on the topic of electricity, conducted by natural science primary school teachers in multigrade settings. Following these observations, video-stimulated recall interviews and individual semi-structured interviews were conducted with each teacher for triangulation purposes. Drawing on this data, the chapter demonstrates that primary science teachers adopt numerous instructional strategies aimed at supporting and enhancing learners' understanding of natural science concepts. The chapter also reveals the significant challenges associated with multigrade teaching, where primary science teachers expressed frustrations with the increased workload and the multifaceted demands placed on them in these environments. Considering these complexities, the chapter emphasises an important role on policy makers and teacher education institutions to embrace multigrade teaching and identify sustainable strategies to support primary school teachers.

S. Zulu (✉)
Science and Technology Department, Wits School of Education, University of the Witwatersrand, Johannesburg, South Africa
e-mail: Sphamandla.zulu1@wits.ac.za

T. Nkambule (✉)
Curriculum Studies and Social Science Department, Wits School of Education, University of the Witwatersrand, Johannesburg, South Africa
e-mail: Thabisile.nkambule@wits.ac.za

D. Sibanda
Science and Technology Department, School of Education Edgewood Campus, University of KwaZulu-Natal, Pietermaritzburg, South Africa
e-mail: Sibandad@ukzn.ac.za

A. Fitzgerald
RMIT University, Melbourne, Australia

S. Earle et al. (eds.), *Primary Science Learning for Children, Teachers, and Communities*, https://doi.org/10.1007/978-981-95-0561-6_11

Keywords Electricity topic · Multigrade · Primary science · Rural schools · Teaching

11.1 Introduction and Background

The education system should be both inclusive and flexible to accommodate the diverse learning needs of learners from varying socioeconomic backgrounds. While adaptability is essential (İşler, 2024; Vally, 2018), it is important to acknowledge that societal inequalities are often reflected and perpetuated within the educational framework. This is particularly evident in the context of multigrade classrooms, which exist to offer education primarily in farm and rural areas characterised by a limited number of school staff, low learner enrolment, and constrained infrastructure and educational resources (DBE, 2010; Du Plessis & Subramanien, 2014; Mulaudzi, 2016). Such settings highlight the challenges of addressing educational disparities, as they often exacerbate the very inequalities the system seeks to overcome. Multigrade classrooms occur when a single classroom contains learners from multiple grade levels, or a single teacher teaches learners from more than one grade in the same classroom (Hyry-Beihammer & Hascher, 2015; Karaçoban et al., 2022; Makoelle & Malindi, 2014; Naparan et al., 2021; The United Nations Educational, Scientific and Cultural Organisation [UNESCO], 1989). Practically, the implementation of multigrade classrooms serves as a contingency strategy employed when a monograde approach is not feasible, thereby ensuring access to education for all learners and address social justice.

There are approximately 5837 multigrade primary schools in South Africa (grades 1–9) from the nine provinces, with majority in the Eastern Cape, KwaZulu-Natal, Mpumalanga, Limpopo and Northwest Provinces. This is an indication that multigrade classrooms will continue to exist in the South African education system, within the context of rationalisation process for unviable schools. Notwithstanding this reality and its challenges, it is still expected of a single teacher with learners from more than one grade in the same classroom, to offer quality and effective teaching, using innovative and creative methodologies (Msimanga, 2020; Thaba-Nkadimene, 2017). Thus, well-trained teachers and continued professional development are important to support quality teaching and learning in complex and dynamic classrooms contexts. In this chapter, we explore rural teachers' practices in primary science multigrade classrooms and examine their science teaching experiences. The chapter was guided by the following research questions:

(a) How do rural primary teachers teach science in multigrade classrooms?
(b) What are the teachers' experiences of teaching science in multigrade classrooms?

11.2 Conceptual Framework

Teaching in multigrade classrooms aims to address issues of social justice, because the intention is to provide quality education for all learners, irrespective of the social and school contexts. To understand teachers' practices in multigrade classrooms, we use Phillips et al.'s (2010) Learning Environment, Learning Processes and Learning Outcomes (LEPO) framework. The three components are interrelated because the activities which are part of learning (Learning Processes) consider the knowledge, behaviours, skills or understanding that need to be demonstrated (Learning Outcomes) and the environment which facilitates the learning (Msimanga, 2020). The authors posit that the learning environment, irrespective of the context, should be designed to enable teaching, learning processes, as well as the assessment of the learning outcomes to determine the achievement of the learning outcomes (Phillips et al., 2010). It was important for us to hear about rural teachers' teaching practices and experiences in the multigrade environment because in rural multigrade classrooms, the learning environment plays a crucial role in fostering engagement and development, especially when considering that South Africa has thousands of multigrade schools. Thus, the LEPO framework 'synergises all aspects of *teaching* (italics added) and learning, and supports new learning environments and methods by integrating learning environment, learning processes and learning outcomes' (Sumanasiri et al., 2015, p. 55), which is appropriate in this study where different grade levels are combined and taught in one classroom.

According to Phillips et al. (2010), the LEPO framework incorporates Biggs' 3-P Model of Presage, Process and Product and Laurillard's Conversational Framework that involves interaction of the teacher and learner, interaction of the learner and the world, adaption of the world by the teacher and learners' action and reflection on learners' performance. However, we only used LEPO and Laurillard's Conversational Framework because they encourage teachers to create a positive learning environment that will permit learners to interact for positive outcomes to be achieved (Sithole, 2017). Some criticisms for the LEPO framework are that it 'does not prescribe how learners and teachers interact with learning environments, processes and outcomes' (Phillips et al., 2010, p. 2501). Similarly, Obiria (2017) noted that LEPO 'does not specify how learners and teachers interact with learning environments, processes and outcomes' (p. 31). The purpose of this study was not to prescribe teachers' practices and nature of interactions with the learners in a multigrade classroom during science lessons, rather to gain insight on their classroom practices and experiences of teaching science in multigrade classrooms. We used two of the three Phillips et al. (2010) components: Learning Environment and Learning Processes and Laurillard's Conversational Framework to observe and analyse the nature of classroom interactions and teachers' actions and reflections. Figure 11.1 below shows the LEPO and Laurillard's Conversational Framework in the context of this study.

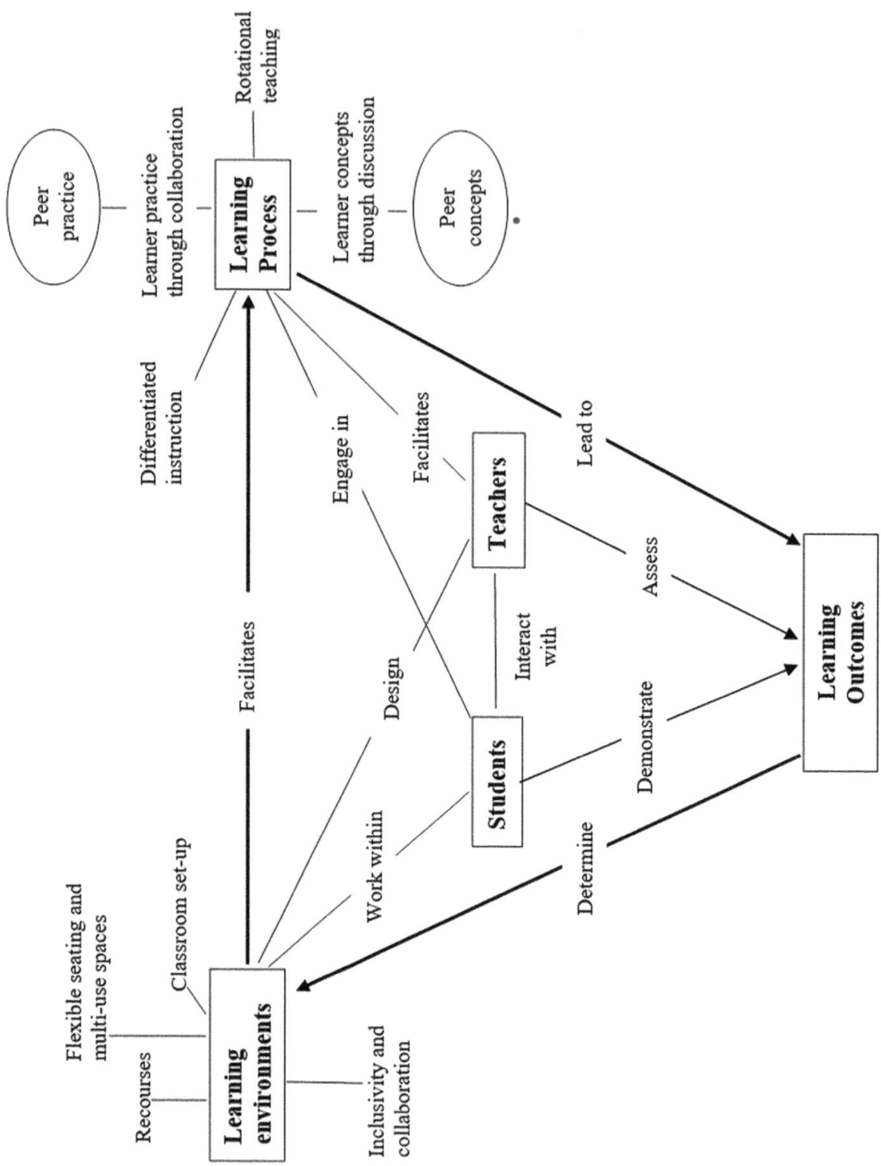

Fig. 11.1 LEPO and conversational frameworks in the context of this study

11.3 Teachers' Ability to Provide Differentiated Instruction

Teachers' ability to provide differentiated instruction to meet the varying needs of their students determines the success of multigrade classes. Differentiated instructions refer to teacher responses to students' learning needs by altering the teaching strategies to suit their learning abilities (Van Geel et al., 2022). A review by Deunk et al. (2018) revealed that implementing differentiated strategies promoted student's achievement, but the use of various strategies was found to be effective only when the instruction was adapted to the students' needs. Therefore, teachers' training backgrounds, professional knowledge base for multigrade teaching (Taole, 2017); multigrade teaching experience and availability of instructional material resources (Ballesteros et al., 2016) among others, have a role to play in the success of multigrade teaching practice. According to Magno (2014) some of the problems experienced by teachers in multigrade teaching are linked to lack of training and insufficient funding. A study conducted by Smeets et al. (2015) in single grade classrooms, show fewer teachers were able to differentiate instructions during the teaching of mathematics. The teaching in multigrade classes is challenging in the sense that the teachers must implement differentiated instructions and in other countries such teachers receive an allowance (Mulryan-Kyne, 2004) but this is not encouraged in South Africa. While providing differentiated instructions, science primary school teachers need to cover grade-specific content as stipulated by the South African Department of Basic Education (DBE) in the curriculum guide. This means that teachers need to be creative on how they allocate time slots to different subjects and between different subject matter knowledge within a subject. The curriculum coverage needs some professionalised knowledge base for teaching in multigrade classes.

From the South African school curriculum, Natural Science (NS) as a subject is officially introduced from Grade 4 until Grade 9. The Grades 4–6 science is taught as a combination of 'Natural Sciences and Technology' and then Grades 7–9 science is taught as 'Natural Science' (DBE, 2011). This means that, Grade 6 science teachers must teach content on Biology (life and living), Chemistry (matter and materials), Physics (energy and change), Astronomy (planet, earth and beyond), and Technology (structures, processing, and systems and controls) in one subject (DBE, 2011). The main concepts taught in Grade 6 under energy and change, which was the interest for this study, include, Electric circuits; Electrical conductors and insulators; and Mains electricity, which is taught over a period of 8 weeks. The main concepts for Grade 5 under energy and change, were stored energy in fuels, energy and electricity, and energy and movement. The multigrade teaching that was observed in this study occurred between grades 5 and 6, and grades 6 and 7 (Grade 6 learners usually range between ages 11–13 years of age). In most South African primary schools, including those worked with in this study, a primary class is taught by one teacher who would teach all the school subjects including NS. Considerations teachers need to make when encouraging discussions and collaborations include peer teaching in their multigrade classrooms.

11.4 Collaborative Peer Learning Opportunities

The opportunities to learn from and assist peer students as demonstrated in the framework in Fig. 11.1, is important, especially for multigrade teaching contexts. Though not common in the South African schools, collaborative learning as a teaching strategy, is used by some rural multigrade classrooms. The strategy allows learners to be more creative when completing tasks (Naparan & Alinsug, 2021). However, a study by Barron (2003) that investigated collaborative interactions amongst primary school learners showed that learners display low-quality in terms of collaboration in classroom activities. Gillies and Boyle (2010) indicated that teachers also face challenges when structuring collaborative activities such as monitoring group work time. Collaborative practices are encouraged by some teachers but opportunities to collaborate are hindered by various factors (Le et al., 2018), hence there was a need for this chapter to find out more about these factors from a multigrade perspective.

11.5 Methodology

11.5.1 Research Design and Procedures

This study adopted a case study research design. In this chapter, we considered themes emerging across the cases of individual teachers, to inform discussion of teaching practice. Case study design was considered necessary in the research context because the boundaries between the studied phenomenon and the context to which it exist are not clearly separated (Yin, 2018). The role of this study then was to unearth learning and teaching experiences of natural science teachers teaching in rural South African multigrade classrooms. The study employed qualitative research approach due to its affordances for in-depth exploration of a situation or process, to gain insightful understanding of that situation (Creswell, 2012).

11.5.2 Data Collection and Analysis

The data collection involved six natural science teachers from six rural primary schools located in one municipality in Newcastle town in KwaZulu-Natal Province, South Africa. All six teachers were teaching multiple grades and ages in one classroom, but we observed and recorded teachers when teaching the Grade 6 content. For this reason, we purposefully selected all teachers. Located within a group of non-probability sampling techniques, purposive sampling was appropriate as we chose participants based on their possession of the characteristics being sought (Cohen et al., 2011). These characteristics being Natural Science teachers, teaching in a Grade 6 rural multigrade classroom and being willing to participate in this study.

We then explained the study to both teachers and learners while seeking permission from both teachers, learners, and parents for classroom observations and interviews with teachers. This was done after getting ethical approval from relevant institutions (University of the Witwatersrand and the KwaZulu-Natal Department of Education). Linked to protection of identity and anonymity, participants were ensured that they would be given pseudonyms in reporting about their data in this study (McMillan & Schumacher, 2010; Scott & Morrison, 2005). Table 11.1 presents the pseudonyms given to teachers and their schools.

The six participant teachers were observed teaching natural science lessons on the theme of Energy and Change, particularly the electricity topic. An observation checklist or protocol was used for all the lesson observations. Following classroom observations, we rewatched the lessons and identified episodes to prompt teachers during the video-stimulated recall interviews (VSRI), which was conducted after lesson observations to allow teachers to reflect on their practice and share their pedagogical reasoning on the observed pedagogical practices. Lastly, we conducted semi-structured interviews with individual teachers. In total, we observed eleven (11) classroom lessons; conducted six VSRI's; and six semi-structured interviews. Also, the classroom observation focussed on Grade 6 classroom teaching. However, some teachers taught Grades 5 (11 years of age) and 6 (12 years of age) in one classroom, Grades 4, 5 and 6 in one classroom, while others taught Grades 6 and 7 (13 years of age) in one classroom. Teacher experiences of teaching in these classrooms were of interest for this study because teachers could narrate, and we also made observations of the contextual conditions under which teachers are teaching. All the participating schools were non-fee-paying, government schools located in rural communities. All

Table 11.1 Selected primary science teachers' biographical information

Teacher	John	Mary	Beauty	Happiness	Albertina	Sipho
Schools	Success primary	Sunshine primary	Hope primary	Progress primary	Star primary	Twilight primary
Gender	Male	Female	Female	Female	Female	Male
TE	9 years	7 years	16 years	4 years	19 years	14 years
Natural Science TE	4 years	2 years	7 Years	4 years	12 years	8 years
Teacher qualification	Bachelor of Education (Inter/Sen)	Diploma in Public Management/ PGCE	ACE	Bachelor of Education (Inter/Sen)	ACE	NPDE
Subject specialisation at university/ college	Mathematics	Management/ Accounting	Natural Science (Inter/ Sen)/Life Sciences	Social Science/ IsiZulu language	Mathematics/ English	Natural Sciences

Note ACE = Advanced Certificate in Education; PG = Postgraduate; NPDE = National Professional Diploma in Education; Inter/Sen = Intermediate and Senior phase

teachers were generalist primary school teachers, meaning they were teaching various subjects in addition to natural science. See Table 11.1 for teachers' biographical information.

The analysis of data for this study focused on VSRI's and semi-structured interviews, although we also draw from the classroom observations. We used inductive coding to organise and analyse the data. The aim of inductive coding was to allow research findings to emerge from the frequent, dominant or significant themes inherent in raw data, without the restraints imposed by structured methodologies (Thomas, 2003). The coding process involved (1) preparation of raw data, making copies of the narratives; (2) close reading of narrative texts of the interviews; (3) code creation, screening and coding teachers' narratives for their reflections on teaching experiences; (4) category creation, which involved a comparison of all coded narrative texts for commonalities, differences, and anomalies; and (5) category revision and refinement, which included linking, deleting, and combing categories to create themes. The results section reports on two themes that emerged from the inductive coding process. The identified themes across the cases were guided by our research questions, as demonstrated in the following section below. Particular attention was given to multigrade environments and how it shaped teachers' pedagogical practices including interactions with learners and learning outcomes.

11.6 Findings

11.6.1 Primary Science Teachers' Pedagogical Practices in Rural Multigrade Classrooms

The importance of demonstrations to support learners' understanding of science concepts was emphasised by the sampled primary school science teachers. These teachers predominantly used authoritative teacher talk and basic demonstrations during the grade 6 natural science lessons. Most of the interviewed primary science school teachers mentioned the importance of demonstration to simplify the theory and practical work to be easy for learners to remember, because demonstration and practical work re-emphasises scientific concepts. The responses below illustrate the points:

> **John**: Demonstration, demonstration and practical for methods, when we do the practicals together, they listen very well. Children like to do practical things, they like to see things happening, you don't have to tell them, if you mix water and sugar, you get a solution, they want to see you mixing water and sugar and see the solution so that they won't forget it. The young ones, they say they have the concrete thinking, which means they want to see things for themselves. It creates a forever picture for learners because if there's a test, if you did it with them in the class, if they did it themselves, they will never forget, they will never fail that thing. I always use practical method, demonstration and practical for methods. I brought coal, the fossils fuel developed millions of years ago, to show them how electricity is made, including the advantages and disadvantages, like causing pollution.

Mary: It's not easy because the time is not enough to teach 3 grades, with Grade 4 coming from foundation phase *[Grades 1 – 3]* you need to give them more time. They must read their textbook then I explain after that, I also ask them questions, and class activities to see whether they understood. Demonstrations, it helps a lot, they don't forget easily if you have demonstrated it (electric circuit), yes, they don't. I showed them and the learners enjoyed a lot if you are demonstrating, they understand better. I first told them about the component and then connected the (electric) circuit so that they will see, this include the symbols, how they work. I showed them this is a wire; this is a battery. Yes, demonstration include drawing a simple circuit, see when the switch is open, and I wanted them to see when it's closed switch.

Beauty: I give them the topic, the small ones (Grade 4), one topic per day, after that I'm going to revise that one topic with them, and then I assess by asking them questions to solve individually and as group. Sometimes we are using practical, I ask them to bring some materials from their homes and then we just do it, it is just the basics, to note what's going on here, what is going on there. They forced me to do those two, demonstration and practical work, so that they (learners) see. So, my planning, there is demonstration and then the project, they need more practical, like a (electric) circuit diagram. Don't tell them as a story, once they watched, you give them the opportunity to go and do it at home for themselves.

Although practical demonstrations and hands-on practical work were important for inclusivity of all learners and collaboration which facilitated the learning process, teachers also used eclectic methods, such as pictures, questions and answers, talk, and chalkboard, taking into consideration the type of learners in a classroom and serious challenges with science resources in the rural context where these teachers are teaching. It was also common for teachers to mention class work activities to identify whether learners have understood the topic, which is expected by the curriculum policy. The learning environments designed by teacher Beauty, for example, allows learners to observe practical processes as a class and encouraged to develop these practical skills individually and collaboratively.

11.6.2 Teachers' Experiences of Teaching Science in Rural Multigrade Classrooms: "It is Difficult and Tough; It Affects Your Passion for Teaching"

Teachers alluded to challenges in teaching science in multigrade classrooms, mostly stating time constraints as a major challenge because they must share time amongst two subjects, time that ordinarily would have been used for teaching one subject. For example, Natural Sciences Grade 5 and Natural Sciences Grade 6 (or grades 6 & 7) were sharing a 60-min period and teachers had to divide the one period time allocation among the two grades in one classroom. For teachers that starts multigrade from Grade 4 to 6 like Mary, for example, they mentioned Grade 4 as the most challenging because learners are coming from Foundation Phase where they did not do science and must dedicate a lot of time familiarising learners with the subject, and this was only unique to Mary who was teaching Grades 4–6 in one classroom. However, due to time constraints teachers are concerned that learners do not receive enough content because they must quickly move to another grade because there is no alignment of content taught in the different grades and the concepts taught in higher grades often a build-up from lower grades. So, there is a need to differentiate the instruction of the different grades in a single class even though there is not enough time to even check whether learners understood concepts, hence classwork activities are strategies for most teachers to keep learners busy and relatively quiet while the teacher engages with another grade. This exposed teachers' ability to set-up their classrooms as conducive learning environments for differentiated instruction where learners engaged in learning processes.

It was, however, a concern that teachers could not recap on previous lessons and/or conclude their lessons well, because they were rushing to teach another grade, which means they are usually against time. Thus, time, as important as it is, is a serious challenge for multigrade teachers, particularly for science subject that needs theory and practical work to be covered simultaneously. Also, teachers mentioned that it is usually difficult to cover the curriculum because of various things that they need to do in constrained time due to multigrading process. Some teachers feel guilty that they are not meeting learners' needs and others mentioned that multigrade classrooms affect their passion for the profession, because of the hard work and potential for burnout. It was concerning to hear other teachers mentioning that multigrade teaching affected their families as well due to fatigue and the multiple assessments they need to set and mark.

Happiness: It's hard, the number of periods is equal, Natural Science (NS) and Technology (Tech), you find that it has 2 periods, it means I must separate, spend 30 minutes this side (Grade 5), another one I spend the other side (Grade 6). If I have 1 period, it means 15 minutes, 15 minutes. Because I must teach both grades, I can't leave class without teaching them. When you give them activities and say Grade 5 you will be doing the activity in groups while I'm teaching Grade 6, they make noise. I try my best way, I try it because I end up having to watch both grades, with Grade 5 and then Grade 6, trying to give them attention. So, I'm working like I'm crazy, that is trauma, I know I'm working like crazy. I'm trying to maintain time.

Albertina: It's better now because I'm only teaching Grade 4 and Grade 5 in one class, I used to teach Grade 4, 5, and 6 in one class and in 1 hour. You end up not having passion to teach, you don't have energy, so, you must have your own strategy. As time goes on, you observe as a teacher that you need to have a strategy to teach multigrade classroom. The timetable says I have two classes for 1 hour, and I will decide that in this 1 hour, I will not teach one class. I will give Grade 5 notes and say paste them in your books and leave them and continue teaching Grade 4 in that 1 hour, explain and give them an activity. I will see where I have a double and come back to Grade 5 to explain the notes they pasted and give them an activity and come back to mark all the work. I try and balance that way to be workable, otherwise you won't cover the curriculum. Another strategy, I pick up a learner or learners in Grade 5 that are clever and explain what they need to teach their peers while I continue teaching the young one (Grade 4), that's how I push my work. I keep an eye and ear on them as they teach and when I see and hear that they are not explaining correctly, I re-explain and come back to the grade (4) I'm teaching. So, you must have a strategy to push the work.

Sipho: (multigrade) they are challenging, they are very challenging because you are taking different grades in one class. It's too much, imagine setting papers for those learners in different levels, not in one subject but must do all the subjects. So, who can do that in a normal sense, teachers are stressed, doing work for 4 teachers but you are 1. In Grades 4, 5, and 6 it's not necessarily about the numbers of learners in multigrade classroom, but the amount of work that the teacher must do. Sometimes I do lecture method because I can't say it's 100%, I do a lot of talking so I can say that it is teacher – centred. I know it's not right, we supposed to communicate with them (learners), I end up talking to myself if they don't give me feedback.

The teachers' experiences in multigrade classrooms illustrate a wide range of challenges, particularly related to time management, student engagement and workload. The science learning environments are often strained due to overcrowding, time constraints, and limited resources. The learning processes differ significantly between teachers, with Albertina showing more flexibility and adaptability, while Happiness and Sipho face significant stress and barriers to interactive teaching. In terms of learning outcomes, the pressure to meet curriculum standards often forces teachers to compromise on quality and depth of student learning. Albertina successfully fosters a more interactive and adaptive environment through peer teaching, while Sipho and Happiness struggle to balance their roles due to overwhelming workloads and classroom dynamics. A key idea that assists Albertina is peer teaching, where students work collaboratively to facilitate the learning process. This idea of peer teaching is also used by Beauty and Happiness, where they allow students to practice through collaboration.

11.6.2.1 Teachers' Perceptions of Multigrade Classrooms

Teachers perceive multigrade classrooms as difficult because it is demanding for teachers and requires various skills. Some teachers strategically pair some grades, for example, Grade 4 and 5 science because similar topics are taught at the same time or term, making it easier to teach two grades as one, though this poses its own challenges on curriculum appropriation and continuity. The other challenges for teachers is classroom management, balancing time to ensure that they finish the curriculum, marking assessments for all the grades they teach on time, because they have big numbers that ranged from 25 to 55 learners per multigrade class. Multigrade used to have small classes, which was one of the reasons for combining two or three grades, however, for the participants in this study, their numbers were higher, a reason it has been difficult to manage classrooms and this negatively impacted their teaching and learning processes. The biggest challenge for all the teachers was lack of and/or insufficient science teaching resources which constrained some learning processes which could have led to positive learning outcomes under normal circumstances.

Albertina: In primary you don't choose what to teach, they just allocate you subjects and say teach these. Multigrade classrooms doesn't want a teacher that is not flexible, who will follow ATP policy as is, you need to understand the kind of learners you have in your class. While it was tough at the beginning, but I like multigrade classrooms now, it makes you to be a thinker and creative as a teacher.

Happiness: I was so surprised when I arrived in class that ahh Grade 6 and Grade 5, I was confused because I really didn't know, I didn't know multigrade classroom, it was my first time. Multigrade is tough for me because there are too many learners. It was tough, really because if you teach multigrade you have to observe, you can't watch Grade 5 only but also Grade 6. I have 22 learners in Grade 5 and 25 in Grade 6 in one class, so my class is overcrowded, I can't even move within the desks. I can't even check whether they are doing the correct work or doing other work when I'm teaching.

Beauty: I will always bring some things from home, but there are some resources I got a while ago. I used to collect them in my old school and put them in my Kit-board, I have got a Kit-board, this is a new school. Matter and materials, I have got a lot of resources for that, the shortage one, especially, it's energy. I have had a lot of problem; I need a lot of resources for that, and I can't find them easily. I bring my own circuit diagram, and I demonstrate in the classroom.

Teachers' perceptions of multigrade classrooms reveal significant challenges, such as overcrowded learning environments (Happiness) and a lack of resources (Beauty), while also highlighting the adaptability and creativity required to thrive in such contexts (Albertina). According to the LEPO framework, multigrade classrooms impact the learning environment by increasing complexity and demanding resourcefulness of the learning environment to help learners comprehend scientific concepts better. Laurillard's Conversational Framework further underscores the importance of dialogue, interaction, and adaptation in these teachers' approaches to dealing with multigrade classrooms. Teachers like Albertina and Beauty demonstrate resilience and adaptation, but constraints like those described by Happiness can hinder effective interaction and learning outcomes.

11.7 Discussion

Multigrade primary science teachers in South Africa utilise practical demonstrations and hands-on practical work to illustrate scientific concepts and processes, particularly in teaching an electricity topic. These pedagogical choices are not merely reactions to the perceived limitations of young learners but are instead rooted in an understanding of how children at this developmental stage learn best through representations such as hands-on, visual methods which are argued to contribute to cognitive learning of children (Hong et al., 2020). Therefore, the representations (resources) used by teachers were meant to facilitate the learning process (Phillips et al., 2010). Beyond these demonstrations, teachers use additional resources for inclusivity of their learners. Such resources include context-specific analogies, pictures, and class reading to manage their multigrade classrooms.

Also, teachers used peer teaching to facilitate learning processes. These approaches reflect a growing body of research, including that of Naparan and Alinsug (2021), which identifies peer teaching as a useful tool in multigrade settings, hence Beauty, Happiness, and Albertina use peer teaching in this study. The use of peer teaching in these classrooms also mirrors the concept of teacher assistants as proposed by Zavvar et al. (2022). In this study, students in one grade acted as peer instructors while the teacher focused on another grade within the same multigrade classroom, as detailed by teacher Albertina. This is an important strategy for multigrade classrooms (Mortazavizadeh et al., 2017). Moreover, the efforts of rural primary science teachers to procure the necessary materials to facilitate learning, even in resource-limited settings, is a testament to their commitment, akin to what was observed by Quejada and Orale (2018) in the Philippines, where teachers sometimes used their own salaries to buy teaching materials to facilitate learning in multigrade classrooms. The persistence of teachers to secure resources for their multigrade classrooms aligns with international calls for equity and access in education, reinforcing that teachers in under-resourced areas are finding creative ways to meet their students' learning needs despite the constraints in the learning environments. For these teachers, student learning is at the centre of their teaching and can be visualised in Fig. 11.2.

The findings from this study further demonstrate that primary science teachers in rural South Africa are implementing pedagogical strategies grounded in educational theory and guided by the Department of Basic Education's policy documents. Despite the well-documented challenges associated with multigrade teaching, as De Borja et al. (2020) also point out, these teachers remain committed to delivering essential science lessons and activities. This commitment underscores the global importance of supporting teachers in multigrade classrooms to ensure that they can continue to provide quality learning processes even under complex learning environments of rural multigrade teaching.

The teachers' reflections on their experiences reveal that teaching in multigrade classrooms can be exceptionally challenging, often leading to feelings of burnout and disillusionment with the profession. A major obstacle is the lack of training and

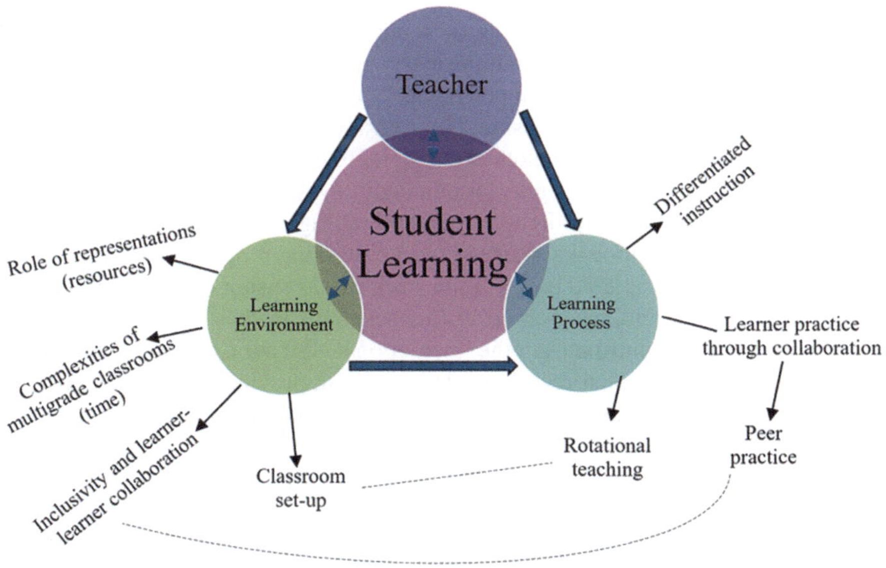

Fig. 11.2 LEPO and conversational frameworks for multigrade primary science teaching

support from their schools and the Department of Basic Education in managing multigrade teaching. Unlike teachers in Lisondra's (2023) study, who benefit from structured lesson plans designed for multigrade classrooms, South African teachers face the added burden of having to create their own resources without formal guidance. The absence of such support hampers the effectiveness of multigrade teaching, underscoring the need for comprehensive professional development programs tailored to these settings, as highlighted by Lisondra.

These challenges are not unique to South Africa. The difficulties experienced by teachers in this study are echoed by De Borja et al. (2020), who reported similar struggles in preparing lesson plans, reproducing learning materials, and navigating curriculum materials that are not suited for multigrade environments. These obstacles hinder teacher productivity and creativity, ultimately affecting student learning process and outcomes. Regrettably, the rural primary science learner suffers as a result. Similar to the findings of Mulryan-Kyne (2004), the current study found that teachers in multigrade classrooms often struggle to effectively implement differentiated instruction, which often led to an increased teacher workload. Kalender and Erdem (2021) confirm that teaching multigrade classes require extensive multitasking, which further compounds the pressures faced by teachers, as also revealed in this study. The issue of insufficient teaching time emerged as one of the most significant challenges for this study, paralleling findings from Zavvar et al. (2022).

11.8 Conclusions

The challenges of multigrade teaching in South Africa are indicative of a broader, global issue. The lack of adequate training and support for teachers in multigrade and multiphase classrooms (such as grades 6 and 7) is contributing to teacher burnout and a loss of passion for the profession. However, despite these challenges, teachers counter the deficit paradigm, developing strategies to manage their classrooms effectively, which offers valuable insights for other educators working in similar contexts worldwide. These findings highlight the need for multigrade teaching to be formally recognized within educational policy and teacher training programs in South Africa and beyond. A well-structured policy framework would ensure that teachers receive the support they need and that their efforts are acknowledged by educational authorities. This study thus contributes to the international education research community by providing evidence-based recommendations for improving multigrade teaching practices, with the potential to inform policy and practice on a global scale.

References

Ballesteros, M., & Ocampo, R. O. (2016). Best practices of multigrade teaching in luna, Apayao, Philippines. *International Journal of Novel Research in Education and Learning, 3*(6), 61–73.

Barron, B. (2003). When smart groups fail. *Journal of the Learning Sciences, 12*, 307–359.

Cohen, L., Manion, L., & Morrison, K. (2011). *Research methods in education* (7th ed.). Routledge.

Creswell, J. W. (2012). *Education research: Planning, conducting, and evaluating quantitative and qualitative research* (4th ed.). Pearson Education Inc.

De Borja, J. M. A., Sigua, E. M. S., & Marasigan, A. C. (2020). Pedagogical practices of multigrade teachers in conducting science learning activities. *IOER International Multidisciplinary Research Journal, 2*(3), 219–228.

Department of Basic Education (DBE). (2010). Rural education. *Rural Education Newsletter, 2*(1), 1–20.

Department of Basic Education (DBE). (2011). *Curriculum assessment policy statement (CAPS), Grades 4–6: Natural sciences and technology*. Government Printer.

Deunk, M. I., Smale-Jacobse, A. E., de Boer, H., Doolaard, S., & Bosker, R. J. (2018). Effective differentiation practices: A systematic review and meta-analysis of studies on the cognitive effects of differentiation practices in primary education. *Educational Research Review, 24*, 31–54. https://doi.org/10.1016/j.edurev.2018.02.002

Du Plessis, A., & Subramanien, B. (2014). Voices of despair: Challenges for multigrade teachers in a rural district in South Africa. *Educational Research for Social Change, 3*(1), 20–36.

Gillies, R., & Boyle, M. (2010). Teachers' reflections on cooperative learning: Issues of implementation. *Teaching and Teacher Education, 26*, 933–940.

Hong, J. C., Ye, J. H., Ho, Y. J., & Ho, H. Y. (2020). Developing an inquiry and hands-on teaching model to guide STEAM lesson planning for kindergarten children. *Journal of Baltic Science Education, 19*(6), 908–922.

Hyry-Beihammer, E. K., & Hascher, T. (2015). Multigrade teaching in primary education as a promising pedagogy for teacher education in Austria and Finland. In *International teacher education: Promising pedagogies (Part C)* (Vol. 22, pp. 89–113). Emerald Group Publishing Limited.

İşler, A. M. (2024). Social policy and cyberfeminism. *Reconstructing Feminism through Cyberfeminism, 275*, 136.

Kalender, B., & Erdem, E. (2021). Challenges faced by classroom teachers in multigrade classrooms: A case study. *Journal of Pedagogical Research, 5*(4), 76–91. https://doi.org/10.33902/jpr.2021473490

Karaçoban, F., & Karakuş, M. (2022). Evaluation of the curriculum of the teaching in the multigrade classrooms course: Participatory evaluation approach. *Pegem Journal of Education and Instruction, 12*(1), 84–99.

Le, H., Janssen, J., & Wubbels, T. (2018). Collaborative learning practices: Teacher and student perceived obstacles to effective student collaboration. *Cambridge Journal of Education, 48*(1), 103–122.

Lisondra, M. J. (2023). Multigrade teachers' lived experiences in hinterland schools: Management plan. *Psychology and Education: A Multidisciplinary Journal, 13*(8), 822–839.

Magno, J. M. (2014). Teaching practices of mathematics teachers in selected public and private elementary schools. *JPAIR Multidisciplinary Research Journal, 17*(1). Retrieved from http://ejournals.ph/form/cite.php?id=12380

Makoelle, T. M., & Malindi, M. J. (2014). Multi-grade teaching and inclusion: Selected cases in the free state province of South Africa. *International Journal of Educational Sciences, 7*(1), 77–86.

McMillan, J. H., & Schumacher, S. (2010). *Research in education: Evidence-based inquiry, MyEducationLab Series*. Pearson.

Mortazavizadeh, S. H., Nili, M. R., Isfahani, A. R. N., & Hassani, M. (2017). Teachers' lived experiences about teaching-learning process in multi-grade classes. *Journal of Education and Learning, 6*(4), 354–363.

Msimanga, M. R. (2020). Teaching and learning in multi-grade classrooms: The LEPO framework. *Africa Education Review, 17*(3), 123–141.

Mulaudzi, M. S. (2016). Challenges experienced by teachers of multi-grade classes in primary schools at Nzhelele East Circuit. *A Thesis Paper Presented to University of South Africa.*

Mulryan-Kyne, C. (2004). Teaching and learning in multigrade classrooms: What teacher say. *Irish Journal of Education, 35*, 5–19. Retrieved from https://bit.ly/3graMLq

Naparan, G. B., & Alinsug, V. G. (2021). Classroom strategies of multigrade teachers. *Social Sciences & Humanities Open, 3*(1), Article 100109. https://doi.org/10.1016/j.ssaho.2021.100109

Naparan, G. B., Ivy Leigh, P., & CastaÃ±eda, M. (2021). Challenges and coping strategies of multigrade teachers. *International Journal of Theory and Application in Elementary and Secondary School Education, 3*(1), 25–34.

Obiria, P. B. (2017). *A location-based privacy preserving model for M-learning adoption to enhance distance education in Kenya.* Unpublished MSc dissertation, Jomo Kenyatta University of Agriculture and Technology, Nairobi.

Phillips, R., McNaught, C., & Kennedy, G. (2010, June). Towards a generalised conceptual framework for learning: the Learning Environment, Learning Processes and Learning Outcomes (LEPO) framework. In *EdMedia + innovate learning* (pp. 2495–2504). Association for the Advancement of Computing in Education (AACE).

Quejada, A. B., & Orale, R. L. (2018). Lived experiences of elementary teachers in a remote school in Samar, Philippines. *Journal Of Academic Research, 3*(3), 1–13. https://jar.ssu.edu.ph/index.php/JAR/article/view/ 7

Scott, D., & Morrison, M. (2005). *Key ideas in educational research.* A&C Black.

Sithole, N. (2017). *Promoting a positive learning environment: School setting investigation.* Unpublished MEd dissertation, University of South Africa, Pretoria.

Smeets, E., Ledoux, G., Regtvoort, A., Felix, C., & Mol Lous, A. (2015). Passende competenties voor passend onderwijs: Onderzoek naar competenties in het basisonderwijs *[Adaptive competencies for inclusive education: Research about competencies in primary school].* ITS.

Sumanasiri, E. G. T., Yajid, M. S. A., & Khatibi, A. (2015). Conceptualizing learning and employability "learning and employability framework". *Journal of Education and Learning, 4*(2), 53–63.

Taole, M. J. (2017). Identifying the professional knowledge base for multi-grade teaching. *Gender and Behaviour, 15*(4), 10419–10434.

Thaba-Nkadimene, K. L. (2017). University of Limpopo student teachers' experiences and reflections during teaching practicum: An experiential learning theory. *International Journal of Educational Sciences, 17*(1–3), 205–214.

Thomas, D. R. (2003). *A general inductive approach for qualitative data analysis.*

UNESCO. (1989). *Multigrade teaching in single teacher primary schools.* UNESCO. https://unesdoc.unesco.org/ark:/48223/pf00000845

Vally, S. (2018). Missed opportunities: The rhetoric and reality of social justice in education. *Koers: Bulletin for Christian Scholarship= Koers: Bulletin vir Christelike Wetenskap, 83*(1), 1–14.

Van Geel, M. V., Keuning, T., & Safar, I. (2022). How teachers develop skills for implementing differentiated instruction: Helpful and hindering factors. *Teaching and Teacher Education: Leadership and Professional Development, 1*, Article 100007. https://doi.org/10.1016/j.tatelp.2022.100007

Yin, R. K. (2018). *Case study research and applications* (Vol. 6). Sage.

Zavvar, T., Abdollahiasl, S., Bahari, P., & Delkhosh, V. (2022). Lived experience of teachers from teaching multi-grade classrooms. *Research in Teaching, 10*(2), 112–143.

Chapter 12
Permaculture in Primary Teaching: Decolonizing and Building a More Socially and Environmentally Responsible World

Roslynn Brain McCann, Kevin J. White, and Liliane Dionne

Abstract Young people continue to experience high rates of loneliness that peaked during the global pandemic, exacerbated by uncertainty about their life purpose alongside climate change anxiety. In these uncertain times, it is essential to empower younger generations with hope and equip them with creative competencies to help create solutions for a better environmental and social future. This chapter will discuss the process of—and research basis for—developing an effective teaching practice based on the Indigenous wisdom of our connection to the Earth, plant and soil ecology, and student empowerment through the lens of the permaculture design framework. Before European arrival on Turtle Island (North America), Indigenous people were taking care of its ecosystems, and with their wisdom, developed an Indigenous scientific method based on symbiotic relationships between plants to ensure food security while maintaining healthy soil ecology. In this chapter, we explore the intersection and complementarity between Indigenous and Western science. We elaborate on the foundation of a plant guild and the integrated, ecological design approach in permaculture. Corn, beans and squash, known as the Three Sisters growing together, are an example of a traditional Haudenosaunee gardening practice that students are invited to explore firsthand in the curriculum. We also present the theoretical background to develop students' agency, and its continuum along this teaching activity. Our pedagogical practice in science, technology and environmental education interweaves Indigenous and Western science, and can be adapted for different areas in North America. Our intention in designing this minds-on, hands-on curriculum is

R. B. McCann (✉)
Utah State University, Moab, USA
e-mail: roslynn.mccann@usu.edu

K. J. White
University of Toronto, Toronto, Canada
e-mail: kevinj.white@utoronto.ca

L. Dionne
University of Ottawa, Ottawa, Canada
e-mail: ldionne@uottawa.ca

S. Earle et al. (eds.), *Primary Science Learning for Children, Teachers, and Communities*, https://doi.org/10.1007/978-981-95-0561-6_12

to support teachers in their efforts to engage and teach young people to understand their co-dependence with Nature.

12.1 Introduction

Western education systems, originating in worldviews separating the self from our environment and each other and embedded within a capitalist framework, are facing surmounting pressures to change (Todd & Robert, 2018). A paradigm shift is required in education, as well as society, for an ecological transition (Zoller, 2013). STEM and environmental education are paramount in helping younger generations be proactive in the climate crisis, and to foster social and environmental justice (Kayumova & Tippins, 2021; Reis, 2021). However, fewer students, especially underrepresented students, are motivated to pursue degrees and careers in STEM and in environmental science globally (Whitcomb & Singh, 2021). According to a US survey by Kennedy et al. (2018), the high level of difficulty of those fields combined with the lack of relevance and interest that many students perceive in STEM and environmental science contributes to their reluctance to pursue these fields. But our society needs these young people to pursue green jobs. One way to increase interest in STEM and environmental education is through a Pedagogy of Liberation where the learning process is driven by reflexivity and creativity (Freire, 2015). The critical thinking embedded in this type of pedagogy is also key to empowering school communities.

As university professors in education, we observe that many pre-service and in-service teachers are lacking confidence in environmental science, climate change, and STEM education. A survey by Field and collaborators in 2020 with 1,231 Canadian K-12 teachers showed that only one-third of teachers indicated feeling knowledgeable enough to teach about climate change. According to this survey, the barriers for teaching climate change include a lack of time, classroom resources, and professional knowledge. If teachers do not feel confident, how can they teach their students in a positive and empowering way? We believe that teachers play a key role in the climate crisis and need support to fully engage in climate science and environmental education, influencing their students to embark on that journey as well. This thinking motivated our creation of a permaculture curriculum[1] for teachers to apply in their classroom as one way to help build a socially and environmentally responsible, action-oriented society. In this chapter, we will describe our collaborative process and how it led to curriculum development.

[1] In Canada, curriculum refers to official K-12 teaching programs; in this chapter however, curriculum refers to an effective pedagogical activity.

12.2 Collaborative Journey as a Team

The seed of our team's collaboration germinated in Spring 2023 during Liliane's sabbatical journey across several American states, in which she engaged with the traditional ecological knowledge used by Indigenous people to keep Mother Earth healthy. Liliane visited Utah State University Moab to establish links with local researchers and met Roslynn, who invited her to a community organic farm for youth. Liliane and Roslynn engaged in community service and shared a community meal at the farm. Because of their shared ecological interests, they collaboratively pursued the development of an environmental education curriculum, especially through the Tableau STE website (Dionne et al., 2019), with a focus on fostering biodiversity, which corresponds to the provincial curriculum of Ontario, Canada. Biodiversity is envisioned through the permaculture design framework, which Roslynn teaches, and the Three Sisters was seen as an ideal Indigenous story to implement in this pedagogical practice.

Due to the Indigenous origins of the Three Sisters story and our desire to respect Indigenous wisdom, we searched for Indigenous scholars from the Haudenosaunee community, which has strong historical ties to Three Sisters gardening. We connected with Kevin, who specializes in Indigenous epistemologies and orality. Given the geographical distance between us (Ottawa/Toronto, Ontario, and Utah), we scheduled monthly Zoom meetings over the course of a year. Between meetings, we worked from a shared Google Drive folder. Liliane and Roslynn often had their non-Indigenous worldviews challenged, and Kevin would share his insights to help them in the process of Indigenization.

12.3 Indigenous Cosmology and a Vision of the Earth We are Intimately Linked with

In our work to develop a curriculum that empowers youth in climate resilience via Environmental Education and STEM, while interweaving Western and Indigenous knowledge, we intentionally created opportunities for young people to learn how they influence and are influenced by nature. The following reflection by Kevin provides insights into this two-way relationship:

> One time, I was feeling grumpy, hurt, and angry with the world. So, I thought a walk in the woods might help. But something strange happened. As I walked deeper into the woods, there was hardly a sound. No birds were singing; no insects were chirping; it was eerily silent. I sat down and tried to calm myself. After a few moments, there was a tentative chirp, a sound from crickets, and life seemed to cautiously try to resume their natural day. The slightest movement by me caused all that natural noise to go silent again. At that moment, I realized how interconnected we all are. It was a very valuable and humbling lesson, but not the end of this lesson. A few days later, I went back onto that very same trail. I was much happier and contented in my mind—what we Haudenosaunee call Good Mindedness. As I was walking, the noisiness of life was teeming all around me; birds were singing, and

> insects were chirping. Then, as I quietly strolled, I came up on a doe, a female deer. She was quietly eating grass by the natural spring—I was so close I could have almost reached out and touched her. Then I stepped on a twig and as it snapped, that doe jumped a long way away from me—startled. She spun around and looked at me, flicked her tail, and headed off deeper into the woods.

The lesson in this is the role we play within ourselves and how we engage with the natural and spiritual worlds around us. Our actions, feelings, and thinking can balance or unbalance creation. This is the delicate balance and how we, as but a part of the natural world, are all interconnected and interdependent. We hope in our teaching that we will leave teachers and students with something to think about and that they will be encouraged to walk in the woods, be grateful, and always leave something for the future generations coming to this island that rests on the back of a great turtle.

As a part of understanding Indigenous worldviews, we learned about the history of the Haudenosaunee Thanksgiving Address. In our meetings, Kevin often shared how Indigenous people cultivate gratitude toward Mother Earth. He told us that gratitude combined with understanding humanity is but one part of honoring the natural world that is highly important to many tribes on Turtle Island, especially the Haudenosaunee who utilize the Thanksgiving Address as a memory tool. From the Thanksgiving Address and shared experiences and memory, Indigenous peoples know that everything is interconnected and dependent on other parts of Creation that exist within a delicate balance. *Ohenten Kariwatekwen* (Thanksgiving Address as we call it today) starts with acknowledging and giving thanks to all parts of creation. As Kevin recalled,

> We begin at the water's edge and move inward to the forest. We give thanks to the people who are gathered. We give thanks to the waters that meet all the needs of creation. We give thanks to the fish, who are not only a source of food for humans but clean the waters too. We give thanks to the plants and trees that provide us with food, shade, and wood to build our homes. We give thanks to the four-legged animals, the birds, and all the animals of the natural world. Sometimes animals are food for humans. Other times the four-legged animals teach us medicines and what plants are good to eat. We also send greetings and thanks to our Grandmother the Moon, as she controls the waters and tides, and guides women in renewing the life of humans on the back of this great turtle. Next, we send our thoughts and gratitude to the stars, our Elder Brother the Sun, and winds. They all play a role in sustaining life on the back of the turtle and for humans. We give thanks and greetings to the Thunderers, who announce the coming rains, and protect humans from Flint's creations that live beneath the earth. We give thanks to the four great teachers who remind us of Good Mindedness, our responsibilities to all of Creation for the bounty in which we exist, and how to get along as diverse humans with each other. We save our best words for the Creator, as it was he who set all of this in motion that allows our continued existence. Finally, we come to the place where we end our words. At this point, we ask others to remember any neglected parts of Creation that we, the speakers, forgot, so that life might continue. Nia:wen (Thank you).

12.4 Braiding Indigenous and Western Science

Strengthening connections between Indigenous and Western Science may positively and creatively impact education systems, which could be richly rewarding for future societies. The Indigenous cosmology presented in the previous section is far from detached, analytical and compartmentalized Western science that focuses primarily on empirical evidence and scientific methodologies. Todd and Robert (2018) discuss how Western knowledge systems isolate the self from nature. In contrast, Indigenous knowledge upholds a holistic perspective in which all living things are interconnected. It emphasizes the sustainable relationships between the environment, community, and spirituality. The link between Indigenous and Western knowledge is complex and multifaceted, as it is influenced by diverse historical, cultural, and social factors. Indigenous knowledge was often marginalized by Europeans through colonization, imperialism, and assimilation policies. This marginalization has resulted in devaluing Indigenous knowledge by Western institutions and systems.

Today, in contrast, there is a growing awareness of the important contributions of Indigenous knowledge systems. The process of decolonization is occurring in many educational institutions in North America. In Canada, following the recommendations of the Truth and Reconciliation Commission from 2008 to 2015, many governmental and educational bodies have been planning and achieving steps towards decolonization. Battiste (2023) defined decolonization as a process of acknowledging the values of Indigenous knowledge, and the wisdom to bring together both Indigenous and non-Indigenous people to learn and evolve with respect for Indigenous knowledge. Our process of developing the curriculum is aligned with this recognition that both knowledge systems can complement each other to offer different perspectives and approaches to understanding the world. There are many advantages to combining both worldviews for students through blending the emotional and cognitive dimensions of knowing the world to find solutions to today's challenges. Indigenous knowledge systems often contain valuable insights into sustainable resource management, biodiversity conservation, and ecological resilience, that can sustain students' agency (Aikenhead & Michell, 2011; Archibald, 2001). By using an Indigenous story in our pedagogical development, we were aware that collaboration between Western and Indigenous knowledge holders requires careful consideration of power dynamics, respect for intellectual property rights, and recognition of Indigenous sovereignty and self-determination (Battiste, 2023). Ethical attitudes, gratitude and informed consent helped build equitable partnerships with mutual benefits (Kim & Leclaire, 2021).

In the following sections, we discuss how transformative pedagogies were used to develop our curriculum, starting from the story-based approach for honoring the ancestral knowledge of Indigenous communities in America, followed by place-based education, which fosters social justice and solutions to environmental and social problems. Nature-based pedagogy centers the curriculum, allowing students to better understand their deep connection to Nature.

12.4.1 *Story-Based Approach: Braiding Indigenous and Western Knowledge*

Recognizing and honoring the contributions of Indigenous knowledge holders is crucial for promoting cultural diversity, environmental sustainability, and social justice, resulting in direct transformations within the day-to-day pedagogy with students (Battiste, 2023). Another advantage of braiding together Indigenous and Western knowledge is to foster cultural revitalization. Many Indigenous communities are working to revitalize and preserve their traditional knowledge systems as a means of asserting cultural identity, self-determination, and sovereignty. This includes efforts to document oral histories, revive traditional practices, incorporate Indigenous language knowledge, and weave transformative pedagogies into formal education systems (Battiste, 2023). In the following sections, we explain how transforming pedagogy helped us to develop the permaculture curriculum. We start with introducing an Indigenous story in the didactic sequence. The story we selected is the Three sisters, told by Robin Wall Kimmerer in Braiding Sweetgrass: Indigenous Wisdom, Scientific Knowledge, and the Teachings of Plants (Kimmerer, 2015). Our practice (see Table 12.1) is intended to help elementary teachers enrich their environmental education repertoire, but also to find ways to decolonize their practices by means of teaching their students with Indigenous stories.

In North America, Indigenous peoples have valued storytelling for centuries as a way of conveying history and knowledge through their families and communities; it has occurred over many generations for sharing and as a foundational way of life (Chan, 2021). Historically, it was left out of Western classroom teaching practice. Today, story-based approaches are experiencing a resurgence as one powerful way of decolonizing education (Davis, 2014).

Table 12.1 Summary of permaculture plant guild lessons for Grade 6

Introductory session	Discuss student gardening experience. Read Three Sisters story and discuss the significance of corn, bean and squash to ensure the survival of Haudenosaunee people and all of humanity
Growing activities	Explore the relationship between the Nitrogen-fixing bean roots and the growth of the corn and squash, while observing the germination and growth of beans
Plant guild planning	Research potential plants for an Apple tree/Haskap berry guild, where the plants work together to enhance each other through their various natural functions, providing physical support, nutrients, and protection, similar to how a healthy forest ecosystem functions
Seed saving	Consider seed banking, food security and food sovereignty. Harvest seeds and create a seed bank to prepare for planting
Corn doll	Make a corn husk doll or similar, to understand the use of various parts of plants in Indigenous communities
Healthy eating	Create a recipe and make soup from the crops grown in a Three Sisters' garden

Storytelling encourages deep learning for students and fosters the exchange of ideas. This approach is essential for anchoring knowledge at the cognitive level, but also at the affective level (Boisselle, 2016). Given that stories are linked to imaginative narratives, they appeal to the imagination and stimulate learners' creativity. Imagination, in turn, leaves room for hypotheses, which are important elements in science and environmental learning. Learning connected to a well-told story promotes more accurate and prolonged retention of the concepts studied, as opposed to learning based on written facts or pictures (Neuhauser, 1993). The approach is often used to preserve and maintain knowledge from the past and to create new visions for the future (Neuhauser, 1993). Jones and Peterson (2017) suggest that teachers should tell stories or invite Indigenous Elders to do storytelling in the classroom to engage and inspire students. He also encourages teachers to ask their students to interpret, tell, write or draw their own version of stories. To develop environmental and climate-change educational practices that include storytelling and Indigenous perspectives, we, as researchers, adopt the *habitus*[2] of contacting communities or knowledge-holders whose stories pertain to develop activities for teachers (Perrenoud, 2001). For example, to enrich an activity about moose *(Alces alces)*, Atisokan, aka Yvon H. Couture, an author who is ethno-historian and writer, kindly provided us with a story about the origin of the animal kingdom, describing how the different animals got their coats, horns and claws to protect themselves against hungry hunters. Abram (as cited in Davis, 2014) believes that our children should be told stories about the land on which they live, stories as close as possible to the Earth, so that they can feel the wind, sun and rain on a deeper level.

12.4.2 Place-Based Education and Pedagogy: Linking Students with Their Community for Social Justice

Another important element of our curriculum design, place-based education (PBE) is an approach to reconnect education with the local environment and community, fostering a deeper, more meaningful learning experience for students (Sobel, 1996, 2004). David Sobel developed PBE to address the disconnect between traditional methods and the developmental needs of children, by providing experiential learning. The first objective in the curriculum we developed is to help students understand what permaculture is, and how this system can help us to nurture both the environment and communities. Permaculture means growing ecological culture; it is a design process based in observation and systems thinking that enables people to create ecologically sustainable and socially just human settlements based in natural patterns and processes (Permaculture Institute, Inc., 2024, Occidental Arts and Ecology

[2] Habitus is a sociological concept first introduced by Pierre Bourdieu, a French sociologist, which refers to the behaviors and preferences of a person or a group, influenced by the social and cultural environment.

Center, 2024). Practitioners and scholars have described permaculture as a catalyst toward promoting transformative pedagogy for education (Luna et al., 2018), fostering learning communities of grassroots practitioners (Ulbrich & Pahl-Wostl, 2019), and offering a holistic integration of the natural world and its relationships into modern ways of life and thought (Aiken, 2017; Lapoutte, 2020). In a qualitative research study with well-known permaculture practitioners (Spangler et al., 2021), permaculture was commonly defined as "an integrated worldview [and a] system-level way of thinking" (p.3).

PBE uses the local community and environment as a starting point to teach concepts in various subjects aiming to immerse students in local heritage, cultures, landscapes, opportunities, and experiences. Several educational movements advocate for the nurturing of a closer link between students and their living environment, and Indigenous territorial pedagogy uses nature as a classroom in culturally meaningful ways (Howard & Kern, 2019). Each territory has its own physical, cultural or historical characteristics, and the planning of teaching and learning interventions needs to take these characteristics into account (Campeau, 2019; Johnson, 2012). Several authors have contributed to theorizing and producing models and methods in PBE adapted to different environments (Somerville, 2011; Zandvliet, 2014). They developed approaches by drawing on the concept of Indigenous territorial pedagogy. Territory brings together the social and ecological dimensions of the human experience, where the environment becomes an active pedagogical agent (Somerville, 2011). Three main principles emerge from Indigenous place-based pedagogy: (1) the body with all its senses as an instrument for reading the environment; (2) the narrative inspired by Indigenous communities who historically owned this territory; and (3) the meaningful experiences from the connections between the person and the environment (Somerville, 2011). The environment, conceptualized as a living entity, is defined by the cultural referents of the people who belong to these lands.

12.4.3 Nature-Based Pedagogy: Developing Students' Connection with Mother Earth

Permaculture provides a framework inviting students to work outside with holistic gardening. We linked this permaculture project with nature-based pedagogy (Barford & Bentsen, 2018) as it invites learners to explore and develop a connection with their local environment while developing their values to see nature as part of themselves and building the knowledge and skills of being and knowing in nature. Because of the multiple cycles experienced in nature, nature-based pedagogy challenges the uniformity of content to be taught, as well as the hierarchies between disciplines (Campeau, 2019). This approach is holistic as it enables students to develop multiple skills, with the help of their peers, the teacher, and the environment itself. It requires us to listen to what the territory can teach us in the permaculture project, which includes respect for nature and learning about the specific characteristics of non-human living species and their habitats (Campeau, 2019). By being involved in the creation of a plant guild in our curriculum, students are provided with opportunities to interact with the surrounding environment, whether through preparing ground, planting, watering, developing pollinator gardens, or harvesting (Barford & Mygind, 2022). By developing a curriculum on permaculture, our objective is to convince teachers and students to actively care about the land by maintaining healthy natural environments, such as by planting the central tree of the guild, which is a constructive response to meeting the United Nations Sustainable Development Goals (SDG). Teachers implementing this curriculum with their students will be caring for the land (SDG 15), preserving water (SDGs 6, 14) and keeping air clean (SDG 13), while fostering the physical, psychological, emotional and social well-being of students (SDG 3).

Nature-based pedagogy was influenced by the Forest School's movement in Scandinavia, mostly in early childhood education and linked to outdoor education, which values children's learning in the forest (McGee et al., 2022). This approach has been gaining popularity in Western countries from K-12. The concept of nature-based pedagogy usually implies prolonged contact between students and the natural environment, which could be constant throughout the day, at a specific time of the day, or on a weekly basis. Students are encouraged to spend time outdoors over many seasons, so that changes and cycles can be observed and experienced (Campeau, 2019). Depending on the location of the school, there are opportunities to bring students into nature; forests are beneficial not only for students learning, but also for their health and well-being (Black, 2020).

Caring for a place such as a garden (the permaculture project), a schoolyard, a pond or an ecosystem near the school nurtures environmental well-being. Nature-based education connects to the concept of Biophilia combined with the notion of Pachamama (Mother Earth). It follows an Indigenous epistemology to develop among students a sense of connectedness with Nature, and to foster environmental well-being (Kellert & Wilson, 1993; Collado-Ruano et al., 2019). In a social context, developing a sense of belonging also includes connectivity with the non-human elements of the environment, such as water bodies, rocks and soil (Birdsall & White, 2020). This link between children (and adults) and nature is essential to finding balance and replenishing their energies.

12.5 The Permaculture Framework: Braiding Indigenous and Western Sciences

The permaculture design framework, within which plant guilds serve as a design tool, provides a model to effectively braid Indigenous and Western knowledge systems for positive change. The framework is based on three ethics (Earth care, people care, and fair share) alongside a set of operational principles ranging from long and slow observation to creatively using and responding to change.

To decolonize science education, it is proposed that a postcolonial science education stance be adopted including Indigenous science that is contextual, community focused, and place-based in recognizing how local populations come to know about the world (Boisselle, 2016). As university professors and teachers, we can develop cultural humility by keeping in mind how the philosophical origins of Western colonial thinking have significantly influenced educational planning (Lekas et al., 2020; Sweet, 2018). In using a permaculture framework, we as curriculum developers allow for a much-needed shift in consciousness among students, from short-term exploitation of people and nature to fostering care for and consideration of the long-term impacts to both (Alderslowe et al., 2018; Ross, 2023). Simultaneously, permaculture entails engaging youth with nature and their respective communities.

In her thesis exploring permaculture curricula for public school children in the United States, Ross (2023) highlights examples of permaculture in primary science education, providing 'detailed lesson plans, broken down into the four seasons, that incorporate permaculture ethics and design principles into gardening activities' (p.34). One curriculum resource, entitled *Outdoor Classrooms: A Handbook for School Gardens* (Nuttall & Millington, 2008), provides comprehensive plans for creating a permaculture-based school garden and a framework for integrating permaculture into K-8 school garden environments and outdoor learning spaces. Another, *Earth Care, People Care and Fair Share in Education: The Children in Permaculture Manual* (Alderslowe et al., 2018), serves as a guide for teaching permaculture ethics and principles to young children in formal and nonformal settings. This manual was designed to help children connect deeply with and see their part in nature, as opposed to perceiving themselves as destructive to the planet (Alderslowe et al., 2018). Lessons are categorized by educational themes ranging from science to home economics. Lastly, the *School Garden Curriculum: An Integrated K-8 Guide for Discovering Science, Ecology, and Whole-Systems Thinking* is another permaculture-based gardening curriculum designed for teachers who are open to transformative pedagogies in their classroom (Christopher, 2019).

12.5.1 Guilds: Patterns in Mutual Support

The detrimental effects of industrial monoculture-agriculture systems on our environment are well documented and range from soil degradation to poisoning our watersheds with synthetic pesticides and fertilizer runoff (Hemenway, 2009; Kimbrell, 2002). Also, as Toby Hemenway cautions in *Gaia's Garden*, 'left unconnected, water, soil, plants, and animals create little more than a collection of fragments' (2009, p.175). Guilds are defined formally as 'a group of plants and animals harmoniously interwoven into a pattern of mutual support, often centered around one major species, that benefits humans while creating habitat' (Hemenway, 2009, p.183). Plant guilds require thinking through relationships between and beyond plants that will ultimately create communities of plants functioning like a healthy ecosystem.

Plant guilds take our thinking beyond monoculture farming and even simply interplanting, which involves combining select plants to avoid negative interactions such as competing for space or light, i.e. densely combining onions, carrots, and lettuce which have different root systems and specific needs (Hemenway, 2009). Instead, guilds entail thinking through both interplanting and companion planting: 'Interplanting combines crops that minimize competition for sun and nutrients. Companion planting blends varieties that enhance each other' (Hemenway, 2009, p.177). Plants exhibit a range of functions that can include fixing nitrogen, attracting beneficial insects, producing mulch, breaking up compacted soil with deep taproots, stabilizing soil with branching roots, and much more. Ultimately, in well-designed plant guilds, our human needs are balanced with the rest of nature. It is our aim to use this permaculture framework to influence educational practice with transformative pedagogies (Barford & Bentsen, 2018).

12.5.2 The Creation of a Curriculum Around a Plant Guild

In realizing the co-benefits of weaving together Western and Indigenous knowledge systems, alongside introducing transformative pedagogies, we settled on developing a practice around biodiversity through the concept of a plant guild (see Figs. 12.1 and 12.2). The activity is intended to immerse students in integrated, systems-level thinking. Students will learn how intentionally selected plants can help fulfil the needs of—and benefit—other plants. In light of growing depression associated with climate change and human destruction of the environment, students can help foster positive ecosystem relationships through plant guilds

12.5.3 Developing Students' Agency and Community Partnership with Schools: A Step Towards the Ecological Transition

Planning and maintaining a plant guild is the root of our curriculum. By being involved in a plant guild project, students will learn to express their agency, and to develop partnerships. The concept of agency is defined as the ability to act according to one's values (Morin et al., 2019). Taking action is central to agency. In the context of this specific curriculum, agency would be the manifestation of independent thinking, with the freedom to choose and to act on one's ideas (Morin et al., 2019). The ecological transformation of school grounds by students is an example of grassroots community organizing that stimulates agency (Black, 2020). To build and maintain such partnerships, networks of students and teachers are organized with people from other entities, such as community groups and local authorities. Projects are set up to solve socio-environmental problems to give students a sense of responsibility.

Goal: Share examples of contributions made to science and technology by First Nations, Metis and Inuit peoples, particularly examples of their relation to Mother Earth and practice of permaculture.

Teaching objectives:
- Explore the history of ongoing Indigenous agricultural practices in the territory where you live.
- Understand the characteristics and uses of select food plants.
- Describe the interdependence between plants, humans, water, soils, and nutrients.
- Examine and taste the benefits of cultivating and eating plants to keep the body and mind healthy.
- Emphasize the importance of plants to society and the environment from a variety of perspectives.
- Suggest personal and collective actions to optimize the positive effects of human activities on plants and to minimize the negative effects.
- Harness student creativity through incorporating art and permaculture.

Overarching Ontario Science and Technology Curriculum: https://www.dcp.edu.gov.on.ca/en/curriculum/science-technology

Curriculum in French: Tableau STE Perte Biodiversité https://tableaust.ca

Learning and teaching approaches: Storytelling, inquiry-based learning and teaching, place-based learning, pedagogy in nature, experiential learning, interdisciplinary approach.

Location: School grounds, community gardens, classroom, community grounds.

Duration: 50 minutes each activity. Possibly one weekly class over a 2-3 month-period.

Materials: Handouts, internet links, vegetables and stock for soup (see recipe), 10-25 corn cob with husks or corn husks, bowl with hot water for husk softening, elastic bands, 1 bag of Earth, organic (ideally Native) seeds.

Working modalities: Plenary, group work, individual work.

Concepts or vocabulary: Permaculture, gardening, plant names, plant anatomy, plant physiology, photosynthesis, nutrients, soil organisms, water conservation, weather, *Haudenosaunee* legends, traditions and vocabulary, *Anishinaabemowin* tradition and vocabulary, Indigenous holistic vision of the world, sustainability, relation to Mother Earth, food security, food sovereignty.

Fig. 12.1 Permaculture plant curriculum overview

In the curriculum, students make soup with the vegetables they grow in the guild. They then share that soup with families in the communities. With such projects, we encourage and enable students to establish fruitful community relationships while also developing their skills to commit to social wellbeing (Jeziorski & Therriault, 2022).

The guild entails improving the environment by creating habitat with trees, shrubs and edible plants. Other ecological transition projects among students, to positively influence their sense of agency efficacy, could involve developing schoolyard wildlife habitats, planting a butterfly garden, installing beehives, or building birdhouses. It could also involve having students develop their teaching competencies by educating other students in the school to adopt environmentally responsible behaviors (Marmot & Bell, 2012).

The development of students' ecological awareness and agency is also correlated with a positive effect on their holistic health (Black, 2020). Thus, by integrating meaningful projects such as permaculture, schools can contribute to developing agency and wellbeing in students, by providing them with learning activities conducive to the manifestation of this power to act by collaborating with the school community and the wider community surrounding the school.

This lesson helps students understand what a plant guild is. A guild is a collection of plants that work together to enhance each other through their various natural functions, providing physical support, nutrients, and protection, similar to how a healthy forest ecosystem functions. Guilds are a design for plant selection and placement in permaculture as their pattern mimics nature's own blueprint for regenerative properties.

Step 1. Draw a basic apple tree* on a white board/flip chart, including the roots.
**For communities in the far North, we suggest you use the Haskap berry (or honeyberry) bush instead.*

Ask the class: "what does this tree/bush provide for us?" (answers: oxygen, fruit, shade, soil stabilization, wind block…let them brainstorm creative answers). Then ask, "what does this tree/bush need to survive?" (answers: carbon dioxide, microbes, water, nitrogen [and other key nutrients such as phosphorus, potassium, magnesium etc – "happy soil"], sunlight, care [pruning, keeping an eye out for nutrient deficiencies, etc]). Alternatively, you can have them work in teams and flipchart to provide their thoughts, then share back to the class. Mention that there are ways we can meet the tree's soil needs, through plant communities or "guilds" (See *Scientific Fact Sheet*). For example, there are what we call nitrogen fixing plants, such as beans (fava, alfalfa, green, runner, field, sweet, peanuts , soybeans, cream, black-eyed or purple-hulled, lupins, lentils, cowpeas, chickpeas); clovers (white, red, crimson, Silver River) and vetches (hairy, American, wood, tufted).

Step 2: As a class, assign each student a potential plant in an apple tree/haskap berry guild:

- Overstory: Apple tree/Haskap berry bush
- Pollinator Attractor: Fennel
- Nutrient Accumulator: Comfrey
- Nitrogen Fixer: Lupine
- Pest Repellant: Perennial garlic, marigold
- Living Mulch: Strawberries, wooly thyme

Ask the students to research their assigned plant - discover some of its benefits and what a benefit could look like if acted out. Once the class is ready, stand the apple tree/haskap berry bush on a chair in the middle of the room, and act out the function selected (i.e. spreading arms wide to provide shade). Then ask who would like to be nearest to the apple tree/haskap berry bush and why, and continue to situate students around and out from the apple tree/haskap berry bush, each acting out their selected function. The class should be fun and chaotic by the end. Once students all return to their seats, discuss the beauty that occurred in such diversity added to the guild.

Step 3: Recap the story of the three sisters. Have students reflect on the functions of the three plants, and how they synchronize together. Then, have students complete their own apple tree/haskap berry bush drawing by selecting some plants that could be in their guild, based on the many layers and functions they could incorporate into one apple "guild", applying their creative energies.

Fig. 12.2 Permaculture plant guild lesson sample

12.6 Discussion and Conclusion

Structural change in education is often a slow and difficult process, yet higher education still falls short in preparing students to engage in systems-level transformation and has much further to go in its own process of decolonization (McCann et al., 2021; Fullan, 2015). Given these institutional inadequacies, transdisciplinary programs and practical applications of systems-level frameworks should be proposed, especially in teachers' education programs. To decolonize and Indigenize their repertoire and learn how to teach generations of students to love the Earth and to substantially decrease their footprint, future teachers need to be adequately prepared. An interdisciplinary space for Indigenous knowledge is required if we want teachers' programs to adopt the 'two-worlds' or 'two-eyed-seeing' approach weaving Indigenous and Western worldviews (Kapyrka & Dockstator, 2012).

In this chapter, we intended to demonstrate that by using transformative pedagogies such as storytelling, place-based, nature-based and agency-based education, and by using permaculture design—as practiced by many Indigenous communities in North America, we can provide pre-service and in-service teachers with an example to initiate decolonization in their practice, and to create sustainability interventions for their classrooms. The aim is to cultivate students' dispositions toward learning about science and the environment and to support the integration of science and sustainability education (Lebo & Eames, 2015). Drawing from the knowledge systems of Indigenous Peoples, who have served as stewards of their lands for millennia, offers valuable lessons for reimagining human Earth relationships. Embedding permaculture principles within education provides a framework for nurturing care and responsibility for the planet while fostering a sustainable worldview and ethic of coexistence within society (Ingram & Maye, 2017).

The adoption of transformative approaches in K-12 is also a pathway to develop awareness and action toward climate change (Bell, 2016). Every step in the classroom to decolonize our methods, using transformative pedagogies to develop students care about the Earth, is a step toward mitigating and adapting to climate change. Our curriculum provides a framework for how teachers can move away from educational instruction embedded in colonialism and separation from nature. Instead, teachers are invited to teach through storytelling, nature and action-based pedagogies, to educate students to view themselves as positive agents for change through active immersion in nature and community. Our interweaving of Indigenous and Western knowledge systems, through the lens of the permaculture design framework, provides a template when it comes to building hope for both our environment and social systems, and a pathway towards climate resilience.

Acknowledgements To Béatrice De Montigny and Marie-Ève Chartrand for trialing and translating the Plant Guilds practice, and to Gefen Bar-On Santor for the English tutoring.

References

Aiken, G. T. (2017). Permaculture and the social design of nature. *Geografiska Annaler: Series B, Human Geography, 99*, 172–191.

Aikenhead, G., & Michell, H. (2011). *Bridging cultures: Indigenous and scientific ways of knowing nature.* Pearson Education.

Alderslowe, L., Amus, G., & Devapriya, D. A. (2018). *Earth care, people care and fair share in education: The children in permaculture manual.* Erasmus +.

Archibald, J. A. (2001). Editorial: Sharing aboriginal knowledge and aboriginal ways of knowing. *Canadian Journal of Native Education, 25*, 1–5.

Barford, K., & Bentsen, P. (2018). Don't ask how outdoor education can be integrated into the school curriculum; ask how the school curriculum can be taught outside the classroom. *Curriculum Perspectives, 38*(2), 151–156. https://doi.org/10.1007/s41297-018-0055-9

Barford, K., & Mygind, E. (2022). Udeskole—Regular teaching outside the classroom. In R. Jucker & J. von Au (Eds.), *High-quality outdoor learning* (pp. 287–297). Springer. https://doi.org/10.1007/978-3-031-04108-2_16

Battiste, M. (2023). *Decolonizing education: Nourishing the learning spirit.* UBC Press.

Bell, D. V. J. (2016). Twenty-first century education: Transformative education for sustainability and responsible citizenship. *Journal of Teacher Education for Sustainability, 18*, 48–56.

Birdsall, S., & White, P. (2020). Making connections: Creating relationships with nonhuman nature, place, and community. *Australian Journal of Environmental Education, 36*, 197–199. https://doi.org/10.1017/aee.2020.35

Black, Y. (2020). 'The play's the thing': A creative collaboration to investigate lived experiences in an urban community garden. *Management Learning, 51*(2), 168–186.

Boisselle, L. N. (2016). *Decolonizing science and science education in a postcolonial space* (Trinidad, a developing Caribbean nation, illustrates). SAGE Open. https://doi.org/10.1177/2158244016635257

Campeau, D. (2019). *Pédagogie autochtone et pédagogie du lieu: démarche hybride pour l'intégration de dimensions culturelles autochtones dans l'enseignement au primaire au Québec.* Thèse de doctorat inédite, Université de Sherbrooke.

Chan, A. (2021). Storytelling, culture and Indigenous methodology. In A. Bainbridge et al. (Eds.), *Discourses, dialogues and diversity in biographical research* (pp. 170–185). Brill Sense. https://library.oapen.org/bitstream/handle/20.500.12657/49770/9789004465916.pdf?sequence=1#page=185

Christopher, K. R. (2019). *School garden curriculum: An integrated K-8 guide for discovering science, ecology, and whole-systems thinking.* New Society Publishers.

Collado-Ruano, J., Madronero-Morillo, M., & Alvarez-Gonzalez, F. (2019). Training transdisciplinary educators: Intercultural learning and regenerative practices in Ecuador. *Studies in Philosophy and Education, 38*(2), 177–194. https://doi.org/10.1007/s11217-019-09652-5

Davis, J. (2014). Towards a further understanding of what Indigenous people have always known: Storytelling as the basis of good pedagogy. *First Nations Perspectives: The Journal of the First Nations Education Resource Centre Inc., 6*, 83–96. www.mfnerc.org/fnp

Dionne, L., Couture, C., Savoie-Zajc, L., & Petringa, N. (2019). The tableau ST project: Inspiring francophone teachers with effective science practices. *Alberta Journal of Educational Research, 65*(4), 360–380. https://doi.org/10.55016/ojs/ajer.v65i4.58447

Field, E., Schwartzberg, P., Berger, P., & Gawron, S. (2020). Climate change education in the Canadian classroom perspectives, teaching practice, and possibilities. *Education Canada,* printemps 2020. https://www.edcan.ca/articles/climate-change-education-canada/

Freire, P. (2015). *Pedagogía del oprimido.* Siglo Veintiuno Editores.

Fullan, M. G. (2015). *The new meaning of educational change* (5th ed.). Teachers College Press.

Hemenway, T. (2009). *Gaia's garden: A guide to home-scale permaculture* (2nd ed.). Chelsea Green Pub Co.

Howard, M., & Kern, A. (2019). The role of story and place in Indigenous science education: Bigfoot in a youth-designed ecological restoration plan. *Cultural Studies of Science Education, 14*, 915–935. https://doi.org/10.1007/s11422-018-9888-7

Ingram, J., & Maye, D. (2017). Niche knowledge systems-challenging or invigorating the AKS? An analysis of the permaculture community in England. In B. Elzen, A. M. Augustyn, M. Barbier, & B. van Mierlo (Eds.), *AgroEcological transitions changes and breakthroughs in the making* (pp. 35–48). Wageningen.

Jeziorski, A., & Therriault, G. (2022). Jalons théoriques et empiriques pour une éducation à l'environnement et au développement durable (EEDD) transformatrice-sociocritique. *Spirale - Revue de recherches en éducation, 2*(70), 101–111. https://doi.org/10.3917/spir.070.0101

Johnson, J. T. (2012). Place-based learning and knowing: Critical pedagogies grounded in Indigeneity. *GeoJournal, 77*, 829–836. https://doi.org/10.1007/s10708-010-9379-1

Jones, M., & Peterson, H. (2017, August 22). *Narrative persuasion and storytelling as climate communication strategies.* Oxford Research Encyclopedia of Climate Science. Retrieved July 24, 2025, from https://oxfordre.com/climatescience/view/10.1093/acrefore/9780190228620.001.0001/acrefore-9780190228620-e-384

Kapyrka, J., & Dockstator, M. (2012). Indigenous knowledges and western knowledges in environmental education: Acknowledging the tensionns for the benefits of a "two-worlds" approach. *Canadian Journal of Environmental Education, 17*, 97–112.

Kayumova, S., & Tippins, D. J. (2021). The quest for sustainable futures: Designing transformative learning spaces with multilingual Black, Brown, and Latinx young people through critical response-ability. *Cultural Studies of Science Education, 16*, 821–839. https://doi.org/10.1007/s11422-021-10030-2

Kellert, S. R., & Wilson, E. (1993). *The biophilia hypothesis.* Island Press.

Kennedy, B., Hefferon, M., & Funk, C. (2018). *Half of Americans think young people don't pursue STEM because it is too hard.* Pew Research Center. https://www.pewresearch.org/short-reads/2018/01/17/half-of-americans-think-young-people-dont-pursue-stem-because-it-is-too-hard/

Kim, E.-J., & Leclaire, S.-L. (2021). To see together without claiming to be another": Stories as relations, against one-direction mode of Indigenous stories traveling. *Griffith Research Online.* https://doi.org/10.15402/esj.v7i1.70002

Kimbrell, A. (2002). *Fatal harvest: The tragedy of industrial agriculture.* Island Press.

Kimmerer, R. W. (2015). *Braiding sweetgrass: Indigenous wisdom, scientific knowledge, and the teachings of plants.* Milkweed Editions.

Lapoutte, A. (2020). The problem is the solution: Can permaculture management regenerate social economy enterprises? *Annals of Public and Cooperative Economics, 91*, 479–492.

Lebo, N., & Eames, C. (2015). Cultivating attitudes and trellising learning: A permaculture approach to science and sustainability education. *Australian Journal of Environmental Education, 31*(1), 46–59. https://doi.org/10.1017/aee.2015.23

Lekas, H.-M., Pahl, K., & Fuller Lewis, C. (2020). Rethinking cultural competence: Shifting to cultural humility. *Health Services Insights, 13.* https://doi.org/10.1177/1178632920970580

Luna, J. M., Dávila, E. R., & Reynoso-Morris, A. (2018). Pedagogy of Permaculture and Food Justice. *Educational Foundations, 31*, 57–85.

Marmot, M., & Bell, R. (2012). Fair society, healthy lives. *Public Health, 126*(Suppl. 1), 4–10.

McCann, R., Spangler, K., & Millison, A. (2021). Life paths to leading systems-level change: Higher education's pitfalls and potential. *Sustainability & Climate Change, 14*(4). https://doi.org/10.1089/scc.2021.0005

McGee, M., Gazdula, J., & Gazdula, E. (2022) Preservation of learning outside the primary classroom: A comparison of policy and practice in Danish udeskole (outdoor school) and English primary Outdoor Education (OE). *Educational Futures, 13*(1) [online]. Available at: https://educationstudies.org.uk/?p=17327. Accessed November 26, 2024.

Morin, E., Therriault, G., & Bader, B. (2019). Le développement du pouvoir agir, l'agentivité et le sentiment d'efficacité personnelle des jeunes face aux problématiques sociales et environnementales: Apports conceptuels pour un agir ensemble. *Éducation et socialisation [Online], 51.* http://journals.openedition.org/edso/5821

Neuhauser, P. (1993). *Corporate legends and lore: The power of storytelling as a management tool.* McGraw-Hill.

Nuttall, C., & Millington, J. L. (2008). *Outdoor classrooms: A handbook for school gardens.* Oregon State University Extension Service.

Occidental Arts and Ecology Center. (2024). *Permaculture design.* https://oaec.org/our-work/projects-and-partnerships/permaculture/

Perrenoud, P. (2001). Le travail sur l'habitus dans la formation des enseignants : analyse des pratiques et prise de conscience. In L. Paquay, M. Altet, E. Charlier & P. Perrenoud (Eds.), *Former des enseignants professionnels. Quelles stratégies ? Quelles compétences ?* (pp. 181–208). De Boeck.

Reis, P. (2021). Desafos à Educação em Ciências em Tempos Conturbados. *Ciência & Educação.* https://doi.org/10.1590/1516-731320210000

Ross, N. (2023). *Permaculture curriculum for public school children in the United States.* Hamline University. https://digitalcommons.hamline.edu/hse_cp/931/

Sobel, D. (1996). *Beyond ecophobia: Reclaiming the heart in Nature education.* Orion.

Sobel, D. (2004). *Place-based education: Connecting classrooms and communities.* Orion.

Spangler, K., McCann, R., & Ferguson, R. (2021). (Re-)Defining permaculture: Perspectives of permaculture teachers and practitioners across the United States. *Sustainability, 13*(10). https://doi.org/10.3390/su13105413

Somerville, M. (2011). Becoming frog: Learning place in primary school. In M. Somerville, B. Davies, K. Power, S. Gannon, & P. de Carteret (Eds.), *Place pedagogy change* (pp. 65–80). Sense Publishers.

Sweet, E. L. (2018). Cultural humility: An open door for planners to locate themselves and Decolonize Planning Theory, education, and Practice. *E-Journal of Public Affairs, 7*(2). https://doi.org/10.21768/ejopa.v7i2.2

The Permaculture Institute, Inc. (2024). *Grow ecological culture.* https://www.permaculture.org/

Todd, K. L., & Robert, V. (2018). Reviving the spirit by making the case for decolonial curricula. In N. N. Wane & K. Todd (Eds.), *Decolonial pedagogy* (pp. 57–72). Palgrave Pivot. https://doi.org/10.1007/978-3-030-01539-8_4

Ulbrich, R., & Pahl-Wostl, C. (2019). The German permaculture community from a community of practice perspective. *Sustainability, 11*, 1241.

Whitcomb, K. M., & Singh, C. (2021). Underrepresented minority students receive lower grades and have higher rates of attrition across STEM disciplines: A sign of inequity? *International Journal of Science Education, 43*(7), 1054–1089. https://doi.org/10.1080/09500693.2021.1900623

Zandvliet, D. B. (2014). Places and spaces: Case studies in the evaluation of postsecondary, place-based learning environments. *Studies in Educational Evaluation, 41*, 18–28.

Zoller, U. (2013). Science, technology, environment, society (STES) literacy for sustainability: What should it take in chem/science education? *Educación Química, 24*(2), 207–221. https://doi.org/10.1016/S0187-893X(13)72464-9

Chapter 13
Exploring Strengths-Based Approaches to Teaching First Nations and Western Sciences

Connie Cirkony, Mijung Kim, Brian Tweed, Jared Carpendale, and Christine Preston

Abstract Worldwide, there are increasing calls to incorporate First Nations perspectives in K-12 science education. Those involved in Initial Teacher Education (ITE) programs in universities have an important role in developing the knowledge, confidence, and capability of current and future teachers to implement culturally responsive approaches in school settings. To support this learning journey, an international group of non-First Nations university science educators from Australia, Canada, and New Zealand formed a community of practice. Together, we share how we are coming to understand First Nations ways of knowing the natural world and explore strength-based approaches in our respective ITE science education programs. In this chapter, we relate stories of our practices through the narrative inquiry method, sharing our experiences and learnings in diverse contexts. Our narratives showcase examples of how we are navigating various policy directives as well as our strategies for implementing curricula, assessment, and pedagogies. Reflecting on our practices, we highlight three key learnings that are relevant for all science educators.

Keywords Culturally responsive science education · Two-Eyed Seeing · Two-ways · Both-ways · First Nations sciences · Indigenous sciences

---

C. Cirkony (✉)
Faculty of Education, Monash University, Melbourne, VIC, Australia
e-mail: connie.cirkony@monash.edu

M. Kim
Faculty of Education, University of Alberta, Edmonton, AB, Canada
e-mail: mijung.kim@ualberta.ca

B. Tweed · J. Carpendale
Te Kura O Te Mātauranga/Institute of Education, Te Kunenga Ki Pūrehuroa/Massey University, Palmerston North, New Zealand
e-mail: b.tweed@massey.ac.nz

J. Carpendale
e-mail: j.carpendale@massey.ac.nz

C. Preston
Sydney School of Education and Social Work, The University of Sydney, Sydney, NSW, Australia
e-mail: christine.preston@sydney.edu.au

S. Earle et al. (eds.), *Primary Science Learning for Children, Teachers, and Communities*, https://doi.org/10.1007/978-981-95-0561-6_13

13.1 Introduction

Worldwide, there are increasing calls to incorporate First Nations perspectives in K-12 science education, including culturally diverse countries such as Australia, Canada, and New Zealand. National policies, including teacher qualification standards, now require educators to understand and include First Nations ways of knowing and being to support reconciliation (e.g., Alberta Education, 2024; Australian Institute for Teaching and School Leadership [AITSL], 2022; Teaching Council of Aotearoa New Zealand, 2024). Science educators must seek ways to bring First Nations perspectives into teaching to enhance learners' appreciation of how people understand the natural world through culturally responsive approaches.

University Initial Teacher Education (ITE) programs play a central role in developing the knowledge, confidence, and capability of current and future teachers to implement culturally responsive approaches in school settings (Morrison et al., 2019). This means post-secondary/tertiary science educators themselves must be culturally competent to confidently and capably implement such approaches.

Most science educators are predominately trained in so-called 'Western modern science' or 'Eurocentric' science and thus have adopted this particular cultural perspective (Aikenhead & Elliott, 2010). Consequently, science educators can benefit from the knowledge that science builds on different bodies of knowledge from other cultures (Cunningham & Williams, 1993), and has been influenced by Western onto-epistemologies (Aikenhead & Ogawa, 2007). Science educators must unpack these perspectives and challenge the assumption that there is one universal way of knowing and doing science (Ogawa, 1995).

To support this learning journey, an international group of non-First Nations science educators from Australia, Canada, and New Zealand formed a community of practice. Together, we share how we are coming to understand First Nations ways of knowing the natural world and explore strength-based approaches in our respective ITE science education programs.

The chapter purpose is to share stories of our practices through the narrative inquiry method, conveying our respective experiences and learnings in diverse contexts to address the following research questions:

(1) How have we incorporated First Nations perspectives into teaching primary science in ITE?
(2) How does our reflection on and discussion about our narratives impact our future practice as ITE science educators?

13.1.1 *About the Authors*

The authors united after an international science education conference presentation forming a Community of Practice (CoP) of university educators from Australia, Canada, and New Zealand. Considered 'settler colonial nations' (Morrison et al.,

2019), these countries have a shared colonial legacy (Hauser et al., 2009), with teaching practices and curricula being largely influenced by Western or Eurocentric philosophies (Blair et al., 2020). We agree that "non-Indigenous educators are critical in influencing change within mainstream society" (Blair et al., 2020, p. 24) and recognise entry to this space as 'outsiders'. Nevertheless, we share a commitment to support current and future teachers to effectively and confidently bridge Indigenous knowledges and ways of knowing into science education (O'Keeffe et al., 2018).

We draw on Beverley and Etienne Wenger-Trayner's (2024, para. 1) conceptualisation of CoP: "Communities of practice are sustained learning partnerships among people who share a concern or a passion for something they do and learn how to do it better as they interact regularly". As a CoP, we aimed to further develop our understandings of First Nations' ways of knowing the natural world, explore culturally responsive approaches, and share our experiences, learnings, resources, and practices, including guidance from our First Nations friends, scholars, and learning communities. Together, we commit to bridging knowledge from Western and First Nations perspectives of science for the benefit of all students (Aikenhead & Elliott, 2010; Morrison et al., 2019), as our shared pathway towards reconciliation.

13.2 Background

We begin by describing culturally responsive and strength-based approaches to embedding First Nations perspectives into a Western view of science education. Culturally responsive teaching is an ideological, ethical, and methodological enterprise concerned with "what to teach, why to teach, how to teach, and to whom to teach with respect to ethnic, racial, cultural, and social diversity" (Gay, 2015, p. 125). This approach ensures voices and ideas are heard from cultures that are often silenced (Snively, 2009) and students have the opportunity to learn more about their own and others' cultures, to prepare them for future community and civic engagement (Gay, 2015). Such aspirations are reflected within professional expectations for teachers (e.g., AITSL, 2022), as well as university, national and global policy directives (e.g., Truth and Reconciliation Commission of Canada, 2012; United Nations General Assembly, 2007; Universities Australia, 2022).

Culturally responsive teaching is premised on the notion that all curriculum and pedagogy are culturally based and has the potential to create "generative and decolonising possibilities" (Morrison et al., 2019, p. 2). First Nations scholars have identified the role of culturally responsive teaching for addressing Indigenous onto-epistemologies in education spaces in Australia (e.g., Nakata, 2011; Rigney, 1999), Canada (e.g., Lewthwaite & McMillan, 2010), and New Zealand (e.g., Bishop et al., 2007). Teaching in this way requires a sustained personal commitment (Morrison et al., 2019), as it challenges educators' dispositions, their practices, understanding of curriculum, and professional relationships (Castagno & Brayboy, 2008). The challenge is to ensure science educators develop their personal and professional capacity to understand and implement culturally responsive approaches relevant to their context.

We now turn to strength-based approaches by exploring the three frameworks which are designed to bring together Indigenous and Western knowledge systems or 'ways of coming to know' in science education. The frameworks are Two-Eyed Seeing, Both-ways/Two-ways, and Kaupapa Māori (Hatcher et al., 2009; Michie et al., 2021; Ober, 2009; Stewart, 2007).

Etuaptmumk or Two-Eyed Seeing (TES) is premised on beliefs that Indigenous and Western scientific ways of knowing are valuable, can co-exist, and inform how we live in the world (Hatcher et al., 2009). TES was developed by Mi'kma'ki Elders Albert and Murdena Marshall in Canada (Bartlett et al., 2012). The Mi'kma'ki word *Etuaptmumk* means 'the gift of multiple perspectives'. As explained by Elder Albert Marshall, TES entails:

> ...learning to see from one eye with the strengths of Indigenous knowledges and ways of knowing, and from the other eye with the strengths of Western knowledges and ways of knowing, and to using both these eyes together, for the benefit of all. (Bartlett et al., 2012, p. 335)

As a strengths-based guiding principle, TES avoids knowledge domination and assimilation by recognising the best from both worlds (Hatcher et al., 2009).

Similar frameworks in Australia include Both-ways and Two-ways learning—terms which are often used interchangeably (Michie et al., 2018). Both-ways was developed by Indigenous scholars such as Yunupiŋu, Marika-Mununggiritj, Harris, Teasdale, and Hughes (Harris, 1990). "Both-ways education is about drawing on and acknowledging skills, language, knowledge, concepts and understandings from both Indigenous and Western knowledge systems" to develop new ways of thinking (Ober, 2009, p. 39).

In Aotearoa New Zealand, Te Tiriti o Waitangi (The Treaty of Waitangi) guides teaching professional standards and practice. Kaupapa Māori theory has developed over the last 40 years to create a space in the academy for research 'by Māori, with Māori and about Māori' (Bevan-Brown, 1998). As developed by Graham Hingangaroa Smith, Kaupapa Māori Theory provides the basis for research centred on Māori worldview, values, language, and practices to achieve Māori aspirations for emancipation (Smith, 2000). This focuses on how Māori can use Western knowledge in conjunction with mātauranga Māori (Māori knowledge) to repower Māori in the twenty-first century (Jackson, 2020). Georgina Tuari Stewart (2010), a prominent Māori researcher in the interface between mātauranga Māori and Western science, also illuminates the political nature of this interface with identity, knowledge, and power circulating within it.

When different knowledge systems are used in parallel, they can offer unique knowledge and innovations informing context-specific decisions (Cajete, 1999). These strength-based frameworks present authentic and compelling possibilities for guiding culturally responsive approaches in primary science education.

13.3 Methods

To share our experiences of embedding First Nations perspectives in our science education programs, we drew on a Narrative Inquiry (NI) approach to inquire into our experiences through constructing stories within "larger cultural, social, and institutional narratives" (Clandinin & Caine, 2008, p. 542). Through this process of constructing and sharing stories, we built and developed trust (Dubnewick et al., 2018), allowing for diverse and scholarly critique through in light of the literature (Clandinin & Caine, 2008).

Following Clandinin and Caine (2008), the authors have generated autobiographical narrative inquiries that foreground our firsthand experiences in relation to our respective cultural, social, and institutional contexts. We drew on data gathered from observations, journal records, interviews, photographs, and anonymous pre-service (i.e., university students in ITE programs) teacher unit evaluations to further substantiate our experiences.

Analysis involved an individual and collective reflective inquiry process into our narrative understandings and stories, while considering our respective cultural, social, and institutional contexts (Clandinin & Caine, 2008). Through this reflective inquiry process, we identified themes emerging across our stories of practice. These themes are addressed in the discussion, where we engage with the research literature and extend our learning as a CoP.

13.4 Narratives

The following five narratives highlight the authors' experiences of embedding First Nations perspectives into science education programs. Each narrative is presented in the original voice of the author, who is identified by their initials. Our first two narratives address how policy directives impact practice.

13.4.1 Science Education in the New Zealand Context (JC)

New Zealand has a unique educational context as it influenced the Te Tiriti o Waitangi, a signed agreement between Māori and the British Crown in 1840. Two versions of Te Tiriti exist: Te Reo Māori (Māori language) and in English. Although likely intended as translations of each other, important differences have led to much debate over the last two centuries.

The signing of Te Tiriti had important policy implications for education in New Zealand during that time. For instance, in 1858, The Native Schools Act was passed, and an extension was again passed in 1867. During this time, Māori communities were offered schools, teachers, and resources by the government. However, the use of Te Reo Māori was actively discouraged (see Seve-Williams, 2013). The effects of such policies meant that the use of Te Reo Māori rapidly decreased and were only beginning to be restored in the second half of the twentieth century. The Māori Language Act (1987) declared Te Reo Māori as an official language of New Zealand. These policies impacted the national science curricula.

Like other countries, New Zealand has a national curriculum that has undergone reform. In 1993, a national curriculum framework was developed and published (see Ministry of Education, 1993), which included a brief section titled "Māori and Science" which focused on making science education accessible for Māori students, with emphasis on acknowledging Māori language and customs. This curriculum was refined in 2007, with more use of Te Reo Māori, and acknowledgment of Te Tiriti (see Ministry of Education, 2007). However, within the science learning area, the focus shifted from students learning about content to understanding the epistemological underpinnings of science, including scientific inquiry processes.

The New Zealand government recently announced a refresh of the national curriculum, for implementation in schools by 2027. A new conceptual framework guides this curriculum, which intends to shift beyond acknowledging Te Tiriti to actioning the principles in this agreement. While in draft form, science education, and changes at the governmental level are still ongoing, this document positions mātauranga Māori alongside other knowledge systems.

ITE for future teachers: Different lenses for worldviews

Our university only offers a graduate qualification for ITE. Pre-service teachers must complete an undergraduate qualification before they complete a one-year graduate program, choosing to endorse as early childhood, primary, or secondary teachers. Our primary pre-service teachers have diverse backgrounds and qualifications—these are often not science.

Early in their primary science education journey, one of the first tasks is to elicit their feelings and confidence towards science, which often reveals their negative experiences or valuing science less than other subjects. This transitions to notions of what is science and what counts as science. We also introduce the idea of mātauranga Māori and science, drawing from available resources to have a critical discussion about their meaning and relationships between them. Pre-service teachers often use the Science Learning Hub, an educational resource website, alongside other governmental websites and resources. Examples frequently help them explore the nuanced differences between mātauranga Māori and science—for instance, karaka berries and harakeke (i.e., plants), and hāngi (i.e., cooking with an earth oven).

One recent activity that seems to be useful to support their thinking involves pre-service teachers constructing two concept maps—one for mātauranga Māori and one for science—both inside circles which they cut out. Pre-service teachers often assume that the circles will be used to form a Venn diagram showing the overlap to highlight similarities and differences. However, once they attach paper between them, they see they have made a pair of glasses. This activity prompts discussion of the idea of looking at phenomena through different lenses and world views, using examples to support their thinking.

Across the year, I try to keep these glasses in mind, challenging pre-service teachers to think about links between the science, curriculum, and mātauranga Māori. They keenly look for these links and include them in meaningful and authentic ways. A recent observation of primary education pre-service teachers is a tendency to be quite open to teaching in ways that connect mātauranga Māori and science.

If we want education to shift in a systematic way, what changes are needed at a policy and school level? Interestingly, much of this will link back to the curriculum changes that are currently taking place—I do wonder how the forthcoming refreshed curriculum will promote the connections between mātauranga Māori and science, and how these will be enacted in schools.

13.4.2 Teacher Standards in the Canadian Education Context (MK)

When the new Teaching Quality Standard (TQS) was announced to develop Education for Reconciliation in the Alberta province in 2018, the inclusion of First Nations, Métis, and Inuit (FNMI) knowledges became mandatory in all subject areas (Alberta Education, 2024; Rheaume et al., 2018). When announced, we in ITE classrooms felt anxious and vulnerable because of the lack of our own understanding, experiences and teaching resources of Indigenous worldview. The mandatory components of TQS legitimised our ongoing conversations and struggles to learn and implement Indigenous perspectives in classrooms.

Along the way, new FNMI resources and pedagogies have been developed for classroom implementation (Danyluk et al., 2023); yet, there remains the issues of resistance and tokenism (Scott & Gani, 2018). How to shift integrating Indigenous perspectives from a TQS mandate to a pedagogy that honours other ways of knowing has been a challenge in my ITE classrooms. Below I introduce a story that I share with pre-service teachers. I learned this story from scholars of indigenous worldview years ago and it significantly changed my thinking and teaching about nature.

Beyond tokenism: A values-based approach to teaching ecology

Ecosystems are often introduced in hierarchical relationships such as producers-consumers or predator-prey. Directly and indirectly, human beings are placed on top or last in food chains as the 'strongest' living beings. Yet, emphasizing the interdependence in living and non-living relationships, Indigenous teaching shares that those on the top of the food chain are the most vulnerable. For instance, in the grass-grasshopper-frog-snake-eagle food chain, the eagle is considered strongest. Yet, an Indigenous view suggests it is the weakest where grass depends on sun, water, and soil; a frog depends on grasshoppers which depend on grass, sun, water, and soil; a snake depends on frogs which depend on grasshoppers, grass, sun, water, and soil, etc. The level of dependency increases toward the top of the food chain making the last one the most vulnerable. Human beings then are the most vulnerable. Thus, humans need to respect and appreciate other beings for their survival. With the inter-dependence and interrelationship of beings with humility, respect, and responsibility, this story often results with pre-service teachers re-thinking the language and perspectives of science that we have learned and have been taught in school science classrooms. We reflect on which knowledge is more valuable for future generations to continue to live in this world.

To prepare my pre-service teachers to implement these ideas, I also share a classroom example that I observed in a public elementary school science class, where the teacher and her Grade 5–6 students were working on forest ecosystems. While introducing the concepts of decomposers, consumers, producers, and non-living components, the teacher introduced Indigenous perspectives on ecosystems. She asked students to think about two different perspectives: European and Indigenous (Fig. 13.1).

The teacher explained to elementary students in class:

> As you are doing this, I want you to keep in mind these different perspectives on nature. In these perspectives, humans are on the top or humans are on the bottom. This one is based on control and this one is based on need. They don't need us; we need all of them… We control lots of them, for our house, water, we hunt, we have ideas for traveling, or technology in advance. These two are very different perspectives. So as you are going through this, these decomposers, consumers, producers are not problematic, they are easy to learn about the elements of this, but when you come to the non-living, I want you to keep in mind there is another way of thinking about those elements in nature.

This classroom episode helps my pre-service teachers to reflect on how Indigenous perspectives and knowledge of ecosystem were becoming part of the science discourse and how this approach could enhance science teaching with values, respect and responsibilities in science classrooms.

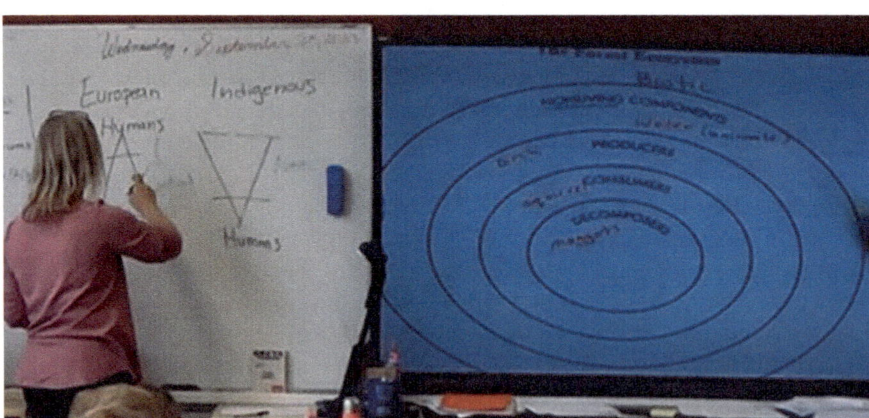

Fig. 13.1 Teacher introducing another way of thinking about nature

Since the new TQS was introduced, my pre-service teachers and I have journeyed together to integrate Indigenous perspectives, openly sharing our challenges, anxieties, and vulnerabilities in the process. Often, we realize how deeply immersed we are in our own worldview, which blinds us to recognizing tokenistic approaches in our practice. We continue to learn the values and possibilities of Indigenous knowledge to overcome the tokenistic approach of the curriculum mandate and ways to honour Indigenous worldview in our classrooms.

Following these first two narratives concerning policies, the final three narratives focus on implementation strategies for content, assessment, and pedagogies.

13.4.3 Revisiting the 'Nature of Science' (CC)

I taught in a graduate B.Ed. Primary program in an Australian university, where pre-service teachers must complete two core science education units. My revision of the 'Introduction to Primary Science' unit was an opportunity to focus on their relationship with science, tapping into their natural affinity for and curiosity about the natural world. In preparation, I undertook my own exploration into the intersections between Western and First Nations sciences (see Cirkony et al., 2023). Writing this paper helped me to build and clarify my own understanding and provided me with a strengths-based blueprint for my approach.

My first challenge came when I started to draft the Nature of Science (NOS) module, which typically highlights a brief history of science along with the methods for practicing science in schools. My past experiences with NOS in school and university textbooks have foregrounded the European scientific revolution in the sixteenth and seventeenth centuries. Through my exploration, I learned that the advent of 'science' as a secular and academy-based discipline took place at the beginning of the nineteenth century, replacing the terms 'natural philosophy' and 'natural history' (Grant, 1997). I was reminded that 'science' was derived from the Latin word "*scientia*", which means 'knowledge' (Cunningham & Williams, 1993, p. 421), as applied to the natural world. While these historical accounts are well established, as are the many great contributions of the scientists of the time, I wondered about the deep-time history of science.

This was my Eureka moment. If we consider science a way to make sense of the natural world, then "science is taken to be as old as humanity itself" (Cunningham & Williams, 1993, p. 408). In other words, since humans have set foot on the planet, we have been making sense of our world in the most profound sense—that is, for continued survival. Though science was built from other knowledge systems, including First Nations knowledge systems, it has been modified and expressed through Western worldviews (Cajete, 1999). I wondered how 'science' was expressed within First Nations knowledge systems and how it might extend our current understanding of the NOS.

Bridging perspectives through common ground and complementarity

While acknowledging that each First Nation has a culturally distinct knowledge system, Cajete (1999, p. 86) noted similarities across these systems that allow for "generalizations to be made regarding Native science". I found Sidney Stephens' (2003) comparison of 'Traditional Native Knowledge' and 'Western Science' an important starting point and included it in my NOS module.

Using common ground to translate between the different systems, there were many shared attributes. For example, both knowledge systems draw from creative, intuitive, rational, and logical thinking, follow systematic empirical approaches, and share their ways of knowing nature through communal social structures (Aikenhead & Michell, 2011). Both systems rely on methods including: "classifying, inferring, questioning, observing, interpreting, predicting, monitoring, problem solving, and adapting" (Johnson et al., 2016, p. 5). Both systems also incorporate technologies to extend their investigations, with Western science using specific tools to "expand scale of direct procedures and indirect observation and measurement" (Stephens, 2003, p. 11). I found these attributes spoke to our inherent disposition toward science as a creative, collective, and systematic sense-making endeavour.

Comparisons between knowledge systems also revealed unique aspects of each. For example, "Respect for all things" (Stephens, 2003, p. 11) reflects the idea of relationality, where all beings and natural elements are interconnected and interdependent. Rather than being separate, humans are considered part of the natural world, although a very *small* part (Hatcher et al., 2009). These ideas complement "mutual reciprocity which simply means a give-and-take relationship with the natural world, and which presupposes a responsibility to care for, sustain, and respect the rights of other living things, plants, animals, and place in which one lives" (Cajete, 1999, p. 79). Both perspectives offer important insights about our place in the world, and our role in 'caring for Country' (i.e., living in sustainable ways), and are foundational for an introductory science education unit.

While these examples are not exhaustive, common ground and complementarity illustrate the value different knowledge systems, how they might work together in beneficial ways (Hatcher et al., 2009; Stephens, 2003). They are a helpful starting point for bridging different knowledge systems. Exploring the intersections between Western and First Nations sciences extend our natural dispositions toward science, and potentially offer a renewed engagement with the natural world that point toward more sustainable ways of living.

13.4.4 Two-Way Approach for Curriculum and Assessment (CP)

The Australian curriculum includes Aboriginal and Torres Strait Islander Histories and Cultures (ATSIC) as a Cross-Curriculum Priority (CCP) to 'offer opportunities to add depth and richness to student learning' of content across learning areas (ACARA,

2024). The ATSIC elaborations, which provide optional content, function as entry points for Two-way science. This narrative involves a small group of pre-service teachers (15% of the cohort) from a final year elective unit—*Science and Technology Specialisation K-6* in the Bachelor of Education (Primary) program.

I provide pre-class readings to help pre-service teachers understand the rationale behind the ATSIC CCP (Sambono, 2021), and consider the Two-ways approach, illustrating its potential for integrating First Nations Peoples' perspectives and knowledges into teaching science (Michie et al., 2021). I commenced my teaching with a provocation:

> The Australian Curriculum: Science elaborations give us suggested ways to incorporate First Nations Science. BUT how do we implement these effectively in the classroom? How do we create learning experiences that showcase Indigenous science alongside Western science in authentic, culturally safe ways and make use of both perspectives to support primary children's understanding of science?

Using spinning tops, I modelled Two-way thinking relating to Year 1 physics: "investigating the push and pull movements of traditional First Nations Australian children's instructive toys" (ACARA, 2024). Pre-service teachers explored different spinning tops (Fig. 13.2). Playing with the toys, we considered similarities and differences, materials, designs, how forces make them work and design elements that created sound. They applied Western scientific concepts to explain how the tops moved.

Pre-service teachers created a top using a sharpened pencil and plasticine, explored how their tops moved and reasoned how they worked. I asked: What natural materials could be used to make a spinning top? Where might they be found? Who would make them? We discussed how First Nations peoples made spinning tops from the hard shell of a nut, beeswax, plaster or volcanic rock, with a stick through a hole in the centre held by "resin and twine" (ACARA, 2019, p. 101). I gave specific examples

Fig. 13.2 Toy spinning tops

such as how the "Yidinjdji Peoples of the Cairns region of far north Queensland, made *bunbuja*, spinning tops from the gourd of the wax gourd plant (*Benincasa hispida*)" (ACARA, 2019, p. 101), and shared images from the Australian Museum collection. We explored other spinning toys that resembled 'koara' (i.e., propellor toy) and 'ngor-go' (i.e., spinning disc suspended on a string) (Edwards, 2008). We discussed how toys and children's games have long been used to convey indigenous knowledge, develop life skills for adulthood (ACARA, 2019), and teach the physics of movement.

Following Michie et al. (2021), to scaffold Two-ways thinking we:

- discussed the benefits of using local natural materials to make instructive toys;
- considered why Indigenous spinning toys vary from place to place, including different materials, ways of applying forces, rule of games and the spinning medium (air, ground);
- analysed how both knowledge traditions (Indigenous and Western) complement each other; and
- noted how making a top and studying construction and operation of traditional tops helped understand how movement relates to size, shape, position, mass distribution and balance.

In discussing how this learning experience emulated the Two-ways approach, the pre-service teachers discovered many entry points in the science syllabus to integrate First Nations perspectives. We considered the potential of Two-ways in their future teaching and how, as science leaders, they might influence their colleagues.

Assessing learning

In one assessment task, pre-service teachers collaboratively developed a digital teaching resource on a science concept for primary students. Using Book Creator ™ (https://bookcreator.com/), they composed an e-book resource incorporating Western and Indigenous science perspectives and resources including practical tasks for children. For example, in astronomy, each aspect of the Sun, Moon and stars, incorporated both Indigenous and Western ways of knowing. Pre-service teachers had co-created culturally responsive resources to support a strengths-based approach to integration of First Nations perspectives into science education in a new way—wow!

Michie et al.'s (2021) article on teaching with epistemic insight made this assessment successful. Using the Two-way approach pushed the pre-service teachers beyond tokenism. They could see how analysing science concepts from both perspectives could help their future students understand how humans think about and make sense of their world. Although I had guided them in this direction, the pre-service teachers demonstrated how Two-ways science could be achieved. They inspired me to take the next step of introducing Two-ways science into my core units.

13.4.5 Facilitating Respectful Conversations (BT)

In New Zealand, discourse in education is framed by Te Tiriti o Waitangi providing a constitution level basis for interactions between Māori and European settlers, their descendants, and recent immigrants. An ethical challenge of working in the interface between Western science and mātauranga Māori is to honour Te Tiriti that guaranteed Māori sovereignty.

I am concerned with what people *do* in the interface between mātauranga and Western knowledges, and what teachers *should* do in this space. A non-Indigenous person has a different set of ethical duties to an Indigenous person working in the interface, and each has further nuanced ethical guidelines to develop depending on their familiarity with science and Indigenous knowledges.

My approach with current and future teachers is personal and conversational. I intend to initiate conversations which repeatedly return the focus to the fundamental natures of both Māori and Western knowledges and Te Tiriti. As the boundaries and limits become clearer, what can and should be done also becomes clearer. People form their own and necessarily different conclusions about what they should do next.

I focus early conversations on how non-Indigenous teachers speak about Indigenous contexts, practices, and knowledge in relation to teaching science. Initially, teachers are focussed on pedagogies of teaching science as defined in national curricula and, if they are pre-service teachers, how they can meet teaching standards. Usually, teachers begin with an instrumental view of mātauranga. They see Māori contexts, practices, and knowledge as things to be used to achieve the goals of the science curriculum. There is a basic assumption of the universality of science which 'is everywhere, in everything and explains everything'. A common, 'go to' context for science teachers is *hāngi*, cooking with an earth oven. This elicits responses like these:

1. "I can use the Māori practice of hāngi to teach students about heat energy."
2. "When Māori do hāngi, they are doing science (whether they know it or not)."
3. "Can we use science to improve the hāngi?"

Such responses provide context for some deeper discussion about ethics and Tiriti relations. Response one positions science as the important knowledge and relegates the Māori practice to a resource for teaching science. Considering response two, Māori people involved in hāngi use their own cultural knowledge to understand what they are doing, not science. Their knowledge determines the nature of the activity not that of the science teacher who observes from the outside. The third response is about crossing a boundary that ought *not* to be crossed—it is not for the science teacher, and their students to redesign an Indigenous practice based on scientific principles. This reconstructs hāngi as a science practice not a Māori one.

Our conversations, highlight a major obstacle to ethical, Tiriti-honouring behaviour in the interface between science and mātauranga. The acceptance that there is a different, non-scientific, yet still valid, way to understand the world requires the non-indigenous science teacher to withdraw and create space for it. There is always much toing and froing here, and careful use and revision of language. Sometimes emotions are triggered, and some confrontational and misleading statements made (e.g., "Are you saying that science is not important?"). Accepting the nuanced position that the scientific view of an entity is not the entity itself, involves serious thinking by many science teachers. Hāngi can be seen in scientific terms and as its own thing derived from within the Māori world.

In conversations, questions about what we can do and say instead, always arise. I suggest making the distinction between the nature of science and the nature of Indigenous knowledges clear at the beginning of any learning and to keep returning to these distinctions regularly and frequently. We can signal when we are crossing boundaries, why, and how we are then taking a different view of things which are not the things themselves. I then invite teachers to reword their previous responses:

- Response one might become: "If we think like a scientist, we can see that heat energy is involved in the Māori practice hāngi."
- Response two might become: "Hāngi is an integral part of te ao Māori [the Māori world], and we can think about it in scientific terms"; and
- Response three might become: Well…perhaps this question just becomes unthinkable—a science-based version of hāngi is no longer a hāngi.

13.5 Discussion

Our five narratives illustrate the complexities of implementing policy to practice in primary teaching settings. Through a NI approach within our CoP, we have engaged in the living, telling, retelling, and reliving of our experiences of embedding First Nations perspectives into our science education programs. We identified common ideas among our experiences and shared how our stories impacted our thinking and potential future practices. In this section, we address the research questions by drawing on aspects of our narratives.

13.5.1 How Have We Incorporated First Nations Perspectives into Teaching Primary Science in Initial Teacher Education?

Our narrative illustrate how we are in various stages of our journey to support our pre-service teachers to understand, appreciate, and apply First Nations perspectives when teaching science. Despite our varied journeys and contexts, we were all challenged by similar epistemic questions, such as: what counts as scientific knowledge? Our respective approaches push us, along with our pre-service teachers, out of our comfort zone of a Western scientific worldview as we engage with these epistemic and ontological questions of science and science education.

The Two-Eyed Seeing or Two-ways approach help guide us in our practice so that pre-service teachers discuss and learn about different perspectives (e.g., CC's common ground and complementarity, CP's assessment task, and JC's glasses). These approaches help us to overcome the hierarchical epistemic structure between Western and First Nations sciences.

Reforms to educational policy, including curricula, that have emphasised and prioritised Indigenous ways of knowing-being-doing have opened up space for us to start our conversation with current and future teachers. For example, within the New Zealand context, acknowledging and actioning the principles of Te Tiriti o Wiatnagi has positioned mātauranga Māori alongside other knowledge systems in curriculum (JC's narrative). Similarly, in Alberta, Canada, the Teacher Quality Standards explicitly emphasise applications of foundational knowledge about First Nations, Métis and Inuit in curriculum and classrooms. Along similar lines, the revised Australian Teaching Standards emphasise the need for a broad knowledge of, understanding of, and respect for Aboriginal and Torres Strait Islander histories, cultures, and languages. Prior to policy, inclusion was voluntary. Since then, there is a growing number of supports available (e.g., professional learning, resources), along with overall reduced resistances and doubt.

At the heart of this process though, is the recognition that pre-service teachers, like us, have varied experiences, understandings, beliefs, and values about science education and Indigenous ways of knowing-being-doing. Thus, our approaches are designed to challenge assumptions and provide opportunities to explore, compare, analyse, and appreciate different worldviews and knowledge systems and how each uniquely contribute to knowledge construction and problem-solving in our respective contexts. Collectively, it is our hope that our pre-service teachers develop their own understandings of First Nations perspectives, such as interdependency and reciprocity, along with the confidence to bridge them with a Western science perspective.

13.5.2 How Does Our Reflection on and Discussion About Our Narratives Impact Our Future Practice as ITE Science Educators?

Through our shared stories and reflections, we examined our own knowledge, beliefs, and practices on what it means to honour First Nations knowledge systems in science classrooms. For example, an excerpt from BT's narrative: "I can use the Māori practice of hāngi to teach students about heat energy" raises a critical question in our teaching when we showcase Indigenous culture. Is the teacher teaching Western science through an Indigenous example? Two-Eyed Seeing, Both-ways and Two-ways approaches suggest that we need to understand, appreciate, and explain the world from different worldviews and acknowledge the strengths of each knowledge system. Yet, we often see First Nations knowledges through the lens of the

Western worldview and replace them with Western scientific terms and concepts. This approach is commonly seen in many teachers' practices, including our own. We teach Western science through First Nations examples, which effectively places the two knowledge systems in a hierarchical structure. Thus, while we endeavour to make paradigmatic shifts, we need to consider how we might be privileging one paradigm over another.

Our narratives and discussions also highlighted the epistemological differences of two knowledge systems. Though there are similarities in both knowledge systems, such as creative, intuitive, rational, thinking, and problem solving, (Aikenhead & Michell, 2011; Johnson et al., 2016; Stephens, 2003), there are also fundamental differences. For example, the Western science methods and knowledge base are universal, whereas as First Nations knowledges tend to be localised (Aikenhead & Michell, 2011). Thus, it is important to understand and acknowledge the role of culture and place on different First Nations knowledge systems.

We further pondered what different worldviews and knowledge systems suggest to us in terms of knowledge, humanity, and values—as well as politics and power—in our respective contexts. While some of these humanistic attributes have recently been included within the Western school science curricula, they are fundamental to First Nations perspectives. Such humanistic perspectives have to potential to engage of traditionally underrepresented groups in science (Klopfer & Aikenhead, 2022), and are critical for addressing some of the complex global problems (Cajete, 1999).

The metaphor of braiding or weaving First Nations science and Western science suggests we teach different ways of knowing-being-doing as legitimate forms of knowledge, rather than one system replacing the other (Snively & Williams, 2018). By acknowledging epistemic differences, we learn how First Nations perspectives can expand our scientific understanding of the natural world.

13.6 Conclusion

This chapter presented stories of our practices on embedding First Nations perspective into our science programs. We highlighted strength-based approaches to implementing policy, curricula, assessment, and pedagogies. Reflecting on our practices through NI, we highlight three key learnings that are relevant for all science educators. First, we questioned what counts as scientific knowledge, whose knowledge is privileged, and how can different perspectives help us understand our natural world. Second, while educational policy and curricula mandates have been critical drivers to engage with First Nations perspectives, the path to bridging or weaving is neither simple nor intuitive and is confounded by the dominant Western scientific paradigm. Third, there remains more work to be done in science education, requiring significant border crossings and supports to do this authentically and respectfully. Hence, there is a need to develop new content, pedagogies, approaches, and resources to support a "multiscience" education (Ogawa, 1995, p. 583), acknowledging the many culturally-based knowledge systems from which science has been built on.

For those wishing to explore strength-based approaches to bridging First Nations and Western sciences, we suggest starting with one of the frameworks to position the knowledge systems as equal, to support transformations in science education.

References

Aikenhead, G. S., & Elliott, D. (2010). An emerging decolonizing science education in Canada. *Canadian Journal of Science, Mathematics & Technology Education, 10*(4), 321–338. https://doi.org/10.1080/14926156.2010.524967

Aikenhead, G. S., & Michell, H. (2011). *Bridging cultures: Scientific and indigenous ways of knowing nature.* Pearson

Aikenhead, G. S., & Ogawa, M. (2007). Indigenous knowledge and science revisited. *Cultural Studies of Science Education, 2*, 539–620. https://link.springer.com/article/10.1007/s11422-007-9067-8

Alberta Education. (2024). *Teaching quality standard.* https://open.alberta.ca/publications/teaching-quality-standard

Australian Curriculum, Assessment and Reporting Authority (ACARA). (2019). *Aboriginal and Torres Strait Islander histories and cultures cross-curriculum priority. Content elaborations and teacher background information for Foundation to Year 6.* https://www.australiancurriculum.edu.au/media/5653/ccp-tbi-f-6-ver5-online.pdf

Australian Curriculum, Assessment and Reporting Authority (ACARA). (2024). *Cross curriculum priority.* https://v9.australiancurriculum.edu.au/f-10-curriculum/f-10-curriculum-overview/cross-curriculum-priorities

Australian Institute for Teaching and School Leadership. (2022). *Australian professional standards for teachers.* https://www.aitsl.edu.au/docs/default-source/national-policy-framework/australian-professional-standards-for-teachers.pdf

Bartlett, C., Marshall, M., & Marshall, A. (2012). Two-Eyed seeing and other lessons learned within a co-learning journey of bringing together indigenous and mainstream knowledges and ways of knowing. *Journal of Environmental Studies and Sciences, 2*(4), 331–340. https://doi.org/10.1007/s13412-012-0086-8

Bevan-Brown, J. (1998). By Māori, for Māori, about Māori - Is that enough? In T. P. Hauora (Ed.), *Te oru rangahau - Māori research and development conference proceedings 7–9 July, 1998* (pp. 231–246). Massey University.

Bishop, R., Berryman, M., Cavanagh, T., & Teddy, L. (2007). *Te Kōtahitanga (Phase 3) Whānaungatanga: Establishing a culturally responsive pedagogy of relations in mainstream secondary school classrooms.* Report to the Ministry of Education. https://thehub.swa.govt.nz/assets/documents/40792_Te_Kotahitanga_Phase3_0.pdf

Blair, N., Stonechild, B., Goulet, L., Cook, B., Rowland, D., Sammel, A., & Whatman, S. (2020). A dialogue around indigenizing education and emerging themes. In A. Sammel, S. Whatman & L. Blue (Ed.), *Indigenizing education* (pp. 3–28). Springer.

Cajete, G. A. (1999). *Native science: Natural laws of interdependence.* Clearlight Publishers.

Castagno, A. E., & Brayboy, B. M. J. (2008). Culturally responsive schooling for Indigenous youth: A review of the literature. *Review of Educational Research, 78*(4), 941–993. https://doi.org/10.3102/0034654308323036

Cirkony, C., Kenny, J., & Zandvliet, D. (2023). A Two-Eyed seeing teaching and learning framework for science education. *Canadian Journal of Science, Mathematics and Technology Education, 23*, 340–364. https://doi.org/10.1007/s42330-023-00276-z

Clandinin, D., & Caine, V. (2008). Narrative inquiry. In L. M. Given (Ed.), *The Sage encyclopedia of qualitative research methods* (pp. 542–545). SAGE. https://doi.org/10.4135/9781412963909.n275

Cunningham, A., & Williams, P. (1993). De-centring the 'big picture': The origins of modern science and the modern origins of science. *The British Journal for the History of Science, 26*(4), 407–432. https://doi.org/10.1017/S0007087400031447

Danyluk, P., Plante, M., & Wessel, S. (2023). Integrating indigenous perspectives into teacher education in Alberta. *Papers on Postsecondary Learning and Teaching, 6*, 28–35.

Dubnewick, M., Clandinin, D. J., Lessard, S., & McHugh, T. L. (2018). The centrality of reflexivity through narrative beginnings: Towards living reconciliation. *Qualitative Inquiry, 24*(6), 413–420. https://doi.org/10.1177/1077800417727762

Edwards, K. (2008). *Yulunga: Traditional indigenous games*. Australian Sports Commission.

Gay, G. (2015). The what, why, and how of culturally responsive teaching: International mandates, challenges, and opportunities. *Multicultural Education Review, 7*(3), 123–139. https://doi.org/10.1080/2005615X.2015.1072079

Grant, E. (1997). History of science: When did modern science begin? *The American Scholar, 66*(1), 105–113.

Harris, S. (1990). *Two-way Aboriginal schooling: Education and cultural survival*. Aboriginal Studies Press.

Hatcher, A., Bartlett, C., Marshall, A., & Marshall, M. (2009). Two-Eyed Seeing in the classroom environment: Concepts, approaches, and challenges. *Canadian Journal of Science, Mathematics and Technology Education, 9*(3), 141–153. https://doi.org/10.1080/14926150903118342

Hauser, V., Howlett, C., & Matthews, C. (2009). The place of indigenous knowledge in tertiary science education: A case study of Canadian practices in Indigenising the curriculum. *The Australian Journal of Indigenous Education, 38*(S1), 46–58. https://doi.org/10.1375/S132601110000082X

Jackson, M. (2020). Where to next? Decolonisation and the stories in the land. In B. Elkington, M. Jackson, R. Kiddle, O. R. Mercier, M. Ross, J. Smeaton, & A. Thomas (Eds.), *Imagining decolonisation* (pp. 133–155). BWB Books.

Johnson, J. T., Howitt, R., Cajete, G., Berkes, F., Louis, R. P., & Kliskey, A. (2016). Weaving indigenous and sustainability sciences to diversify our methods. *Sustainability Science, 11*, 1–11. https://doi.org/10.1007/s11625-015-0349-x

Klopfer, L. E., & Aikenhead, G. S. (2022). Humanistic science education: The history of science and other relevant contexts. *Science Education, 106*(3), 490–504. https://doi.org/10.1002/sce.21700

Lewthwaite, B. E., & McMillan, B. (2010). 'She can bother me, and that's because she cares': What Inuit students say about teaching and their learning. *Canadian Journal of Education, 33*(1), 140–175.

Māori Language Act. (1987). https://www.legislation.govt.nz/act/public/1987/0176/latest/whole.html

Michie, M., Hogue, M., & Rioux, J. (2018). The application of both-ways and two-eyed seeing pedagogy: Reflections on engaging and teaching science to post-secondary Indigenous students. *Research in Science Education, 48*(6), 1205–1220. https://doi.org/10.1007/s11165-018-9775-y

Michie, M., Rioux, J., & Hogue, M. (2021). Incorporating Two-ways thinking about time into the science curriculum. *Teaching Science, 67*(1), 36–43.

Ministry of Education. (1993). *Science in the New Zealand curriculum*. Learning Media.

Ministry of Education. (2007). The New Zealand curriculum. *Learning Media*.

Morrison, A., Rigney, L. I., Hattam, R., & Diplock, A. (2019). *Toward an Australian culturally responsive pedagogy: A narrative review of the literature*. University of South Australia. https://apo.org.au/node/262951

Nakata, M. (2011). Pathways for indigenous education in the Australian curriculum framework. *The Australian Journal of Indigenous Education, 40*, 1–8. https://doi.org/10.1375/ajie.40.1

Ober, R. (2009). Both-ways: Learning from yesterday, celebrating today, strengthening tomorrow. *The Australian Journal of Indigenous Education, 38*(S1), 34–39. https://doi.org/10.1375/S1326011100000806

Ogawa, M. (1995). Science education in a multiscience perspective. *Science Education, 79*, 583–593. https://doi.org/10.1002/sce.3730790507

O'Keeffe, L., Paige, K., & Osborne, S. (2018). Getting started; exploring pre-service teachers' confidence and knowledge of Aboriginal culture. *Asia Pacific Journal of Teacher Education, 47*(2), 152–175. https://doi.org/10.1080/1359866X.2018.1531386

Rheaume, J., Spencer, B. L., Donlevy, J. K., Gereluk, D., & Brandon, J. (2018). Alberta's new teaching quality standard and its implications for teacher education. *The Journal of Educational Thought, 51*(2), 162–183.

Rigney, L.-I. (1999). Internationalization of an indigenous anticolonial cultural critique of research methodologies: A guide to Indigenist research methodology and its principles. *Wicazo Sa Review, 14*(2), 109–121. https://doi.org/10.2307/1409555

Sambono, J. (2021). The Aboriginal and Torres Strait Islander histories and cultures cross-curriculum priority: Cultural responsiveness in science education. *SASTA Journal, 1*, 4–13.

Scott, D., & Gani, R. (2018). Examining social studies teachers' resistances towards teaching Aboriginal perspectives: The case of Alberta. *Diaspora, Indigenous, and Minority Education, 12*(4), 167–181. https://doi.org/10.1080/15595692.2018.1497969

Seve-Williams, N. (2013). Equality of educational opportunity, merit and the New Zealand education system. *Current Issues in Language Planning, 14*(2), 244–253.

Smith, G. H. (2000). Māori education: Revolution and transformative action. *Canadian Journal of Native Education, 24*(1), 57–72.

Snively, G. (2009). Money from the sea: A cross-cultural indigenous science activity. *Green Teacher, 86*, 33–38.

Snively, G., & Williams, W. (Eds.). (2018). *Knowing home: Braiding indigenous science with western science, Book 2*. https://ecampusontario.pressbooks.pub/knowinghome2/

Stephens, S. (2003). *Handbook for culturally responsive science curriculum*. Alaska Native Knowledge Network. https://files.eric.ed.gov/fulltext/ED451986.pdf

Stewart, G. M. (2007). *Kaupapa Māori Science* (Thesis, Doctor of Philosophy (PhD)). The University of Waikato, Hamilton, New Zealand. https://hdl.handle.net/10289/2598

Stewart, G. M. (2010). *Good science? The growing gap between power and education*. Sense Publishers.

Teaching Council of Aotearoa New Zealand. (2024). *Our code, our standards*. https://teachingcouncil.nz/professional-practice/our-code-our-standards/

Truth and Reconciliation Commission of Canada. (2012). *Truth and reconciliation commission of Canada: Calls to action*. https://www2.gov.bc.ca/assets/gov/british-columbians-our-governments/Indigenous-people/aboriginal-peoples-documents/calls_to_action_english2.pdf

United Nations General Assembly. (2007). *United Nations declaration on the rights of indigenous peoples, A/RES/61/295*. http://www.unhcr.org/refworld/docid/471355a82.html

Universities Australia. (2022). *Indigenous strategy annual report*. https://universitiesaustralia.edu.au/wp-content/uploads/2022/08/UA_Indigenous_Strategy_Annual_Report_May-2022.pdf

Wenger-Trayner, B., & Wenger-Trayner, E. (2024, March 7). *Communities of practices*. https://www.wenger-trayner.com/communities-of-practice/

Chapter 14
Using an Online Community of Practice to Support Geographically Isolated Early Career Primary Science Specialists in Initial Teacher Education to Develop Their Professional Practice

Carolyn Julie Swanson and Karyn Saunders

Abstract This chapter discusses how an online Community of Practice [CoP] supported the development of Early Career Academics [ECAs] in primary science education. The CoP was formed to mitigate geographic isolation and the paucity of domain specific professional development. The study participants were five ECAs and the CoP facilitator who were New Zealand primary science education specialists. High quality pre-service science teaching is important as it effects graduate teacher's confidence and efficacy in teaching primary science. Data from semi-structured interviews was thematically analysed using themes derived from Wenger-Trayner and Wenger-Trayner's (2015) CoP components of community, domain and practice. Findings showed that creating a warm environment that fostered authentic conversations was crucial in the success of the CoP. The connections generated within and from the CoP allowed participants to build an identity as a science academic. The facilitated conversations allowed participants to develop domain specific understandings, research capabilities and aspirations and hone their teaching practice. Upon reflection, it was recognised there was a stronger emphasis on research development. A need to include a deeper focus on teaching practice in these conversations was identified as a way to promote consistent practice across universities and enhance the experiences of pre-service teachers.

Keywords Community of Practice · Early Career Academics · Initial Teacher Education · Primary science · Science education

---

C. J. Swanson (✉)
School of Education, Auckland University of Technology, Auckland, New Zealand
e-mail: carolyn.swanson@aut.ac.nz

K. Saunders
Victoria University of Wellington, Wellington, New Zealand
e-mail: karyn.saunders@trinityschools.nz

S. Earle et al. (eds.), *Primary Science Learning for Children, Teachers, and Communities*, https://doi.org/10.1007/978-981-95-0561-6_14

14.1 Rationale

Ensuring school students demonstrate a solid understanding of science concepts and are scientifically literate concerns governments nationally and internationally (Stewart et al., 2024). School teachers are responsible for ensuring students learn science; so, science education academics in Initial Teacher Education [ITE] need to make certain pre-service teachers can teach science effectively to children in their charge (Weinberg et al., 2021). Concerningly, anecdotal evidence shows that science education academics in ITE in Aotearoa New Zealand's universities often work in isolation due to the relatively low number of academics working in primary science teacher education. This means there is limited scope for professional support and development in the primary science education domain within individual universities. These limitations in professional development may impact the capacity of ITE science education academics to develop pre-service teachers' confidence and proficiency in teaching primary science. This isolation can be more problematic for early career academics [ECAs] as they seek to manage the competing demands of research, teaching and employment precarity (Hollywood et al., 2020). This chapter will outline research into how ITE primary science education academics can be supported during the ECA phase through the formation of an online Community of Practice (CoP). The study was undertaken to answer the following research question:

> How did participating in this online CoP support our development as ECAs in the primary science domain in ITE?

14.2 Literature

This section will outline pertinent literature describing the ITE environment in Aotearoa New Zealand, challenges facing early career academics in ITE in science education, and the mechanism chosen to support their development in this study—joining an online Community of Practice [CoP].

14.2.1 Undertaking Initial Teacher Education in Aotearoa New Zealand

While there are a number of different ITE providers—25 as of December 2023 (Teaching Council of Aotearoa New Zealand, 2023)—this study will focus on those in the university sector. Seven of the eight universities in New Zealand offer primary ITE education. Each university is approved by the Teaching Council of Aotearoa New Zealand [TCANZ] (2023) to deliver different programmes, ranging from bachelor's undergraduate degrees to graduate, postgraduate or master's qualifications.

Students undertaking an ITE qualification must meet the requirements for a university degree at either undergraduate or graduate level and the required Standards for the Teaching Profession (Education Council New Zealand I Matatu Aotearoa [ECNZ], 2017) before obtaining their Provisional Registration. Each ITE programme must undertake regular reviews to ensure that they are delivering the approved programme and are supporting graduates to achieve the teaching standards and complete the programme competently (Teaching Council New Zealand I Matatu Aotearoa, 2019, p. 62). The provisionally registered teachers (PRT) must be capable of stepping into their first teaching role and be responsive to future challenges (Teaching Council of Aotearoa New Zealand, 2022). Once employed as a teacher they undertake a two-year induction and mentoring programme before obtaining a Full Practising certificate if they meet the requisite requirements from the Teaching Council (ECNZ, 2017; Education Review Office Te Tari Arotake Matauranga [ERO], 2024).

ITE providers must ensure each curricular area is covered in their teaching. In primary ITE programmes, this would normally be one academic paper or course, with allotted teaching time varying from 22 to 48 hours. During those hours ITE science education lecturers need to present all aspects of the science curriculum, pedagogical approaches, assessment, planning, resources and connection to Mātauranga Māori. Several of the ITE providers are also integrating curricular areas (e.g. science and technology) so teaching hours in science may be less than stated. Therefore, our primary pre-service teachers may receive very little science instruction before they start classroom teaching. This lack of experience and/or confidence is borne out by 32% of recent graduates surveyed by the ERO (2024) stating they were “unprepared to teach science” (p. 4). However, by the time new graduates have taught for one year, their confidence in teaching science has improved with only “13% percent unconfident” (p. 99). This lack of confidence in science in the school setting may be due to a focus on the “core subjects, such as maths, reading and writing” (p. 100). This scenario is common, as is a lack of professional development in science education for primary school teachers (Swanson, 2021).

14.2.2 Exploring Complexity: Challenges of Being an Academic in ITE

Generally, to become a lecturer (the lowest academic grade) in ITE in New Zealand, universities require a doctoral degree or to be close to completion at the time of appointment. There is no clear route to becoming a teacher educator in ITE. Some come as teacher educators without a PhD and others enter with a research background and may not have an educational background, which means they have different professional development needs (Lunenberg et al.’s, 2014; MacPhail et al., 2014). Many who transition to ITE from a teaching background still strongly identify as ‘teachers’ whilst those who come via a research background may either identify with their discipline or as an intellectual (van Lankveld et al., 2017).

Working in ITE as a teacher educator brings additional complexity. We take Lunenberg et al.'s (2014, p. 5) definition of teacher educators as "*all those who teach or coach (student) teachers with the aim of supporting their professional development*". The authors suggested that teacher educators work in a variety of roles, such as being a teacher of teachers; a coach; curriculum developer; researcher; and gatekeeper. Dickerson et al.'s (2021) enlarged definition included providing professional development for teachers, collaborating with stakeholders, contributing to discourse on the teaching profession and teacher education, and ensuring students receive exemplary education. Working as an academic in science education in ITE requires negotiation of multiple roles, such as teacher educator, curriculum developer, and academic researcher. This dance between roles can be demanding, stressful and isolating (Patton & Parker, 2017).

The early career stage of academia is complex with many competing pressures. New appointees identify feeling isolated, stressed, insecure in their teaching or research skills, and uncertain whether academia is for them (Mulholland et al., 2023; Rienties & Hosein, 2020). Factors such as having unrealistic expectations of the job, poor mentoring and support, and the necessity to prove oneself in all areas, especially in research to obtain permanent employment may have a detrimental effect on the health of ECAs, their academic success and career longevity (Hollywood et al., 2020; Price et al., 2015).

Similar to Price et al. (2015), the term Early Career Academic [ECA] is used rather than Early Career Researcher [ECR], as it represents the multifaceted roles undertaken in the early stages of an academic career. However, Sutherland et al.'s (2013) definition of an Early Career Academic as being "within the first seven years of their first permanent academic appointment" (p. 8) is used as it is placed within the New Zealand context, rather than five years, common in other countries like Australia (Price et al., 2015). ECAs may be employed in one of three areas: predominantly research, predominantly teaching or in a mixed research and teaching capacity (Rienties & Hosein, 2020). While both aspects of the work of an academic are important, academic demands may mean research is prioritised above teaching (Mula et al., 2022). Transitioning securely into an academic role may be supported by attending work mandated professional development but is also impacted by external connections, such as partners and peers at other institutions (Rienties & Hosein, 2020). MacPhail et al. (2014) asserts that higher education-based teacher educators want professional learning to enhance both research skills and educational practice. In the past, many teacher training institutes had less focus on research than on teaching. However, there is a clear expectation that all academics including ECAs meet key indicators in research, teaching and service to their schools and professional communities (Mulholland et al., 2023; Sutherland et al., 2013). This means there can be tension between one's identity as a teacher and one's identity as a researcher, the latter which can be perceived to have greater value in academia (van Lankveld et al., 2017).

14.2.3 Supporting the Transition into Academia

There are several ways that the transition from teacher to teacher educator can be supported. This may include being coached or mentored by a colleague; learning from and with colleagues; undertaking a course in higher education, studying their own practice or joining a CoP (Lunenberg et al.'s, 2014; MacPhail et al., 2014; Mulholland et al., 2023). Professional learning on teaching provided for new employees may be too generic or pedagogically simplistic to meet the needs of teacher educators, who already have considerable expertise in teaching (MacPhail et al., 2014). Working in science education adds another layer. Weinberg et al. (2021) asserts that teacher educators who work in STEM must address five components of their identity—"teacher educator, STEM educator, STEM disciplinary expert, STEM education researcher, and teacher education researcher" (p. S102). To that mix of complexity, we must also adhere to meeting not only the universities assessment requirements but also ensure that our students meet the Teaching Council of Aotearoa New Zealand's (Teaching Council of Aotearoa New Zealand, 2022) professional requirements to be fit for practice.

14.2.4 Community of Practice: A Lens to Explore Learning for Primary Science ECAs

Community of Practice (CoP) will be used in this chapter as a lens to discuss supporting the transition from being a teacher to an academic and supporting the development of research and teaching practice in the primary science education domain.

Wenger (2018) initially described a CoP as a conceptual framework examining how learning and knowing occurs in social settings. He identified four interconnected components—community, identity, practice and meaning. Participants in a CoP develop a sense of *identity* through belonging to and developing their connection with the *community*, as they talk about their *practices* and actions in the domain, making *meaning* both individually and collectively. A later iteration by Wenger-Trayner and Wenger-Trayner (2015) tightened the theory to three components—domain, community and practice. Broadly, a CoP can be defined as a group of people with mutual interests and expertise in a domain who regularly meet to learn, explore and develop their practice (McDonald et al., 2012; Mercieca, 2017; Wenger-Trayner & Wenger-Trayner, 2015). Connections can be enhanced between the members of the CoP through "*boundary objects* - artifacts, documents [and] concepts" and "*brokering* – connections provided by people who can introduce elements of practice into another" (Wenger, 1998, p. 105).

According to McDonald et al. (2012), CoP structures range from self-driven organic structures with minimal support to CoPs that are nurtured and supported with some facilitation to CoPs intentionally created by management as a short-term project to explore predetermined issues or themes in which participation is encouraged. Constructing a CoP may be difficult if participants have differing experiences or expectations, as it takes time to build trust within the community and negotiate a shared vision and practice (Adie et al., 2017). However, there is value in having a range of experiences and academic positions, as they can offer different perspectives and identify potential opportunities which less experienced participants may not recognise (Camarao & Din, 2023). However, senior academics (e.g., Associate Professor, full Professors) are less likely to engage in CoPs than lecturers and senior lecturers citing workload and not meeting their needs (Wilson et al., 2020). It must be noted that ECAs may find it easier to ask questions, share their practice and clarify expectations with those at a similar level than discuss their challenges with those whose status is higher (Mulholland et al., 2023; Rienties & Hosein, 2020).

CoP has been used to examine higher educational settings (McDonald & Cater-Steel, 2017). For example, CoP has been used to enhance interdisciplinary collaboration (Pharo et al., 2014), support pre-service teachers (de Laat et al., 2017), teacher educators (Adie et al., 2017; Patton & Parker, 2017), and ECAs (Cox, 2013; Kensington-Miller, 2021; Pharo et al., 2014; Rienties & Hosein, 2020; Sutherland et al., 2013). Each CoP differs according to members' needs. Some focus on supporting members to develop their practice to support students' learning (Camarao & Din, 2023; Kensington-Miller, 2021) or exploring practice and research capacities (see for example, Patton & Parker, 2017).

14.3 Methodology

This section outlines the participants and setting, data collection, and analysis.

14.3.1 Participants and Setting

There were six participants in this study—five members of an online CoP and the facilitator. To blur their identities, information will not be specifically attached to a pseudonym. However, please see Table 14.1 for general information. All participants had a PhD and worked within the primary science education community. Five were employed on permanent contracts at New Zealand universities. The sixth was an independent researcher, employed part-time in a primary school. The facilitator was employed in a predominately research role during data collection. The four members working in ITE had been in their roles for less than seven years and were employed in combined research and teaching roles. Two were senior lecturers and two were lecturers.

An online science community was started by the facilitator in 2022. It was termed an online CoP as members were a group of ECAs with mutual interests and expertise in primary science education working in ITE, who regularly met to discuss their

Table 14.1 Description of the online CoP members

Category	Differentiation	Number of participants
Years since PhD obtained	Less than seven years	Five
	Greater than seven years	One
Employment at time of research	University	Five
	Primary school (part-time)	One
Job title	Lecturer	Two
	Senior lecturer	Three
	Primary science specialist teacher	One
Employment composition	Combined teaching, research and service role	Four
	Predominantly research and service role	One
	Independent researcher	One
ITE teaching focus	Primary and secondary	Two
	Primary only	Two
	Not applicable	Two

practice in the domain to learn with and from each other (McDonald et al., 2012; Mercieca, 2017; Wenger-Trayner & Wenger-Trayner, 2015).

The six participants met monthly through the online video conferencing platform Zoom, for an hour. Attendance was optional but calendared with participants attending as able. The meetings were loosely chaired by the facilitator. The usual pattern was that the participants were asked what they were doing in terms of research and/or practice in their pre-service primary science classrooms. Occasionally there was an invited guest, who would speak on a topic like how to manage the early career stage. Discussion was held about the science educational field and wider educational environment nationally and internationally. The meetings were not formal, and no minutes were taken. The online group also shared life with new babies being shown and support given in navigating university challenges.

14.3.2 Data Collection

Ethics approval No. 0000031242 was obtained through Victoria University of Wellington to explore the dynamics of the online CoP. All participants were interviewed by two researchers who were also members of the online community. The research was undertaken 12 months after the CoP commenced. Semi-structured interviews were chosen to generate data (Cohen et al., 2018). There were eleven questions to springboard discussion, with two examples given below.

On a scale of 1–10, 1 being the lowest and 10 being the highest, how do much do you feel that membership to the ECR group informs your ITE practice? Why have you assigned that number from the scale?
In what way(s) has being a member of the ECR group assisted you to connect with people outside of the group in the field of science education?

The interviews were audio-recorded through Microsoft Teams along with the automatic transcription of the interview.

14.3.3 Data Analysis

An initial stage of familiarisation occurred as the two researchers discussed the interview data after the interviews. After the Microsoft Teams transcriptions were checked for errors, the researchers searched the transcripts for codes inductively and deductively. An adaptation of Wenger's model (2018) of community, identity, practice and meaning making was used to categorise the data segments, which were highlighted and collated.[1] However, further data analysis collapsed the themes into three themes, more closely resembling Wenger-Trayner and Wenger-Trayner's (2015) categories of community, domain and practice. The final themes were invited to belong to the science education community; developing an academic identity within the science education domain; and developing one's practice, which had two sub-themes—expanding our research practice and expanding our teaching practice.

14.4 Findings

In this section the findings will be discussed under the three themes.

14.4.1 Invited to Belong to the Science Education Community

The community in this online CoP was purposively built by the facilitator who brokered connections. They had "existing relationships" with four participants who were approached in May 2022. The initial email was sent to six people with a heading entitled "Emerging science education researcher". The rationale for the group was outlined succinctly. "I'm keen to ensure that you all 'meet' each other, as our new and emerging research community within science education in Aotearoa" (personal communication, May 18, 2022). Of the six people identified in the initial email, one never attended, and one attended once. The other four became the main cohort.

[1] Please note as there were only six participants and as they are known within the science ITE community in New Zealand (and beyond), quotes will not be attached to specified participants.

An ECA in science education at another university was identified at a conference, approached by email and joined the online community in late 2022. The facilitator later explained their rationale, "New Zealand had a group of people who were all forging your way, …[at] slightly different stages, but similar parts of the journey and the science education landscape seemed really important to be able to support." The reason for an online group rather than face-to-face was due to the wide geographical distances between participants, ranging from 134 to 1078 km.

The major reason for joining the online community was to connect "with other ECRs [Early Career Researchers]". It was more than a general chat online but a purposeful "connect[ion] in the work we are doing, and essentially nerd out". They wanted to have "lunchtime catch up online". In fact, one participant noted that "when this meeting clashes with other things, typically I'll prioritise this meeting … to catch up with people, find out what going on … and network".

Belonging to the Emerging Science Education Researcher community was embedded in warm personal relationships. It felt to one participant as if "we're sitting around with a cup of coffee". They further developed the idea noting, "I feel safe with all of the members of the group on a personal level. I love catching up." Another highlighted that the CoP members "care about you on a deeply personal level …. It's like we are a family". The importance of "being able to share those personal things [means] you just develop this sort of deep connection with people". It is the deep connection that is crucial in providing a safe platform to "say what you think and draw on your professional knowledge" in the online discussions.

14.4.2 Developing an Academic Identity Within the Science Education Domain

An identity as a science education academic was built through connections within and beyond the community. The CoP provided a place for building an identity as a science academic. Five of the six participants considered that the group provided a sense of 'professional belonging' between seven and 10 when they self-rated themselves during the interviews. The sense of belonging was dependent upon their individual context. For example, the participant who was not working at a university and with part-time research contracts said they felt "a bit like an imposter" and placed their sense of belonging as a seven. Similarly, another participant who had not been in the field long gave it a five, noting "I'm not sure I'm always talking about the academic stuff … I don't know how much rigour I would bring." In contrast, more experienced participants rated it higher. One noted that "New Zealand is a professional isolation place because of the small numbers of people in science education [in ITE]". They further commented, "I do feel the sense of belonging … the work we all do in science education". A more experienced participant stated, "Belonging to the ECR group does provide me with an identity in science education", in contrast to work where "I am one of the team". They clarified that, "their professional identity is very much wrapped up in science education." Another participant stated they had "gone to an eight from probably a six at the start, which would have been me finding my place." Of importance, they mentioned that through "being part of this group I have an identity as someone in the science field" and have been "invited to a celebration [with] a room full of scientists". The facilitator noted even though they were, "no longer in the ECR space … I feel very much part of the group".

A participant comment identified that negotiating an identity as "an academic is hard work, and I think the group makes it less hard work". "Having a place to share experiences such as the cuts at all of the universities", makes it easier as others understand what you are going through. "Discussing things like PBRF [Performance-based Research Fund] brings a sense of belonging." This community provided a place for the participants to "be encouraged and built up" and "provide a professional mirror and professional support … I feel less] like an imposter at work".

The online CoP also provided members with opportunities to develop their academic self by "connecting with people outside the group", with the community acting as a "conduit to reach other people and [raise awareness] of research opportunities". Members felt it was "easier to connect with someone when you say I'm a member of the community". They also indicated that being "interconnected with NZARE [New Zealand Association for Research in Education]", was also an important way of staying relevant in the field. Talking to "giants of the field" in New Zealand or with invited Australian guests, embeds us into the community and enhances our knowledge of what "people in the community are thinking about science education". Connecting with others outside of the CoP was also important for future career development as it impacts on one's "status growth … in [New Zealand], Australasia … [and] internationally".

14.4.3 Developing One's Practice

Wenger et al. (2002) definition of practice being "a set of frameworks, ideas, tools, information, styles, language, stories and documents that community members share" (p. 29), was used to explore how we expand our specific knowledge about practice in the role of a science ITE academic.

Taking a broader perspective, we develop our practice by the stories we tell and our interactions within the community. The community provided space for the participants to discuss "cool thing[s] and ideas they have" or in the words of another participant, "nerd out" in science. "Confidence and inspiration" in practice is gained through "hear[ing] about what ideas they might have and things they are applying for and conferences and publication". Practice will be explored through two subthemes—expanding our research practice and expanding our teaching practice.

14.4.3.1 Expanding Our Research Practice

There was a stronger focus on the research aspect of being an ECA with discussions "centring more on research and academic citizenship" with members being "supported in our research trajectories". This included discussing things like "Performance Based Research Fund", "brainstorming ideas of what to research", and "ethical processes at other universities". Other members highlighted the usefulness of "professional conversations about what's going on in the world of science education" and to "brainstorm ideas of what to research". Members found "being able to talk the work that I'm doing and hear about the work that other people are doing just really sort of fosters that sense of professionalism". Another member agreed, that "being able to bounce ideas … [and] get critical feedback" was crucial.

Moreover, talking about research practice highlights "what conferences are around", giving impetus to attend as we know other CoP members will be there. This is important as due to geographical distance not all members had met each other. It is wider than talking about our aspirations or activities, we also help each other. This may be through "reviewing an article", "looking at an abstract" or providing a copy of a paper that "was taken offline".

14.4.3.2 Expanding Our Teaching Practice

The data showed thinking about teaching practice was not as strong as the research component, which was not surprising as the CoP's focus was on supporting members' research capabilities. However, the CoP was a "a safe place to talk through some of the challenges" affecting the science community in Aotearoa New Zealand such as "Mātauranga Māori (Māori knowledge) and the resources that are available" and the ramifications of the "curriculum refresh"[2] and possible changes under a new government.

We also explored our practice in "terms of what we are giving to our community [our pre-service teachers], our external communities [such as the schools]…[and asking whether] they approve of what we're teaching." We explored pre-service teachers needs as many were struggling with "getting into classes …. student numbers… or how do we engage them [in learning]?" We acknowledged that "the context in which we are working shapes" how learning is delivered, with pre-service teachers learning through a variety of modes (such as face-to-face, distance, or mixed modes), which intensifies the complexity of teaching and learning.

The value of discussing teaching practice was dependent upon the experience of members and how the teaching courses were structured. An experienced member stated that "a lot of the planning has already been done…. And in terms of the science education stuff, I've got what I want to do and why I want to do it". They

[2] In 2023 the science curricular area of the New Zealand curriculum was being refreshed. Only one draft was released for review before the focus changed due to a new government taking office. Please see Stewart et al. (2024) for commentary on the science curriculum debates.

also noted they had "limited time" to teach all necessary science. Less experienced members highlighted the usefulness of "connecting on our teaching" and finding out what other people's programmes looked like" in terms of "assessments", "readings … and curriculum integration". Viewing both perspectives, the CoP adds to practice obliquely, by providing space to "share things and talk about the different activities", and as members "think about things" discussed after the sessions, inform their practice.

14.5 Discussion

This section will discuss how the findings answered the main research question:

> How did participating in this online CoP support our development as ECAs in the primary science education domain in ITE?

It will be examined by discussing the development of our identity as a science education academic in ITE, our connections to the Science Education domain and our research and teaching practice.

14.5.1 Identity as a Science Education Academic in ITE

The findings showed that belonging to the community allowed members to actively develop an identity as a science education academic in ITE. Wenger (1998) considers that our identity forms in the nexus between "our investment in various forms of belonging and our ability to negotiate the meanings that matter in those contexts" (p. 188). Our belonging to the community was enabled by discussion guided by the facilitator with time allotted in each session to share about the 'cool things' they were doing in research. Having opportunities to share what they were doing "facilitated feelings of competence" (Hardy et al., 2022, p. 168); legitimising each member's work and provided a way to establish an identity (Pyrko et al., 2017, p. 391) as someone in the science education domain. Mapping one's experiences as similar to others in the CoP allowed members to see they belonged to the 'social' community of science education, even if they were not yet experts. Belonging to this online CoP enabled two of the participants to diminish their feelings of inadequacy, which is a risk factor for leaving academia (Hollywood et al., 2020; Mulholland et al., 2023). Belonging as a passive member was not enough, there was desire to grow and take more risks such as 'applying for conferences' with the aim of aligning our practice and research contributions with the science education community (Wenger, 1998). However, this risk taking was optimised through the "social support" that belonging to a CoP provided (Rienties & Hosein, 2020, p. 4).

The CoP provided a forum for the ECAs to be authentic and discuss their aspirations and challenges in the research and teaching space but also personally. Attending the meetings regularly allowed for relationships to be built and provided an "emotional connection" between the members, which Hardy et al. (2022), identifies as an important marker in transitioning from being a teacher practitioner to an academic. Attending online meetings provided a safe space to be oneself, whether that was a 'mother' or a 'science nerd' without fear of being found out as an imposter, which enabled the members' anxiety to lessen (Mulholland et al., 2023). As the members grew more confident, they shared their challenges, including those common to ECAs such as negotiating university systems, research expectations, teaching loads and being inadequate (Mulholland et al., 2023). Having someone to listen, empathise and provide advice made academia "less hard".

14.5.2 Connecting to the Science Education Domain

The findings show that belonging to this community enabled members to connect with each other and with the wider national and international primary science education community. It mitigated the geographical isolation of members who were at tertiary institutes without a large science education department. As the internal connections across the domain of science education were weak within their universities, the CoP members created stronger external connections nationally (Rienties & Hosein, 2020) with each other through the interface provided by the CoP. This interface provided members with "academic, professional and social support" (p. 4) to widen their perspectives about science education and grow their practice.

The facilitator acted as a 'broker' (Wenger, 1998) between CoP members. Similarly, CoP members also acted as brokers between their science education contacts in the wider science education community and each other. Some brokering was intentional, for example inviting a colleague from a previous university to talk to CoP members. Other brokering instances involved 'translating' for members how to negotiate the complexities of applying for conferences or hone their research agenda. How valuable the members found the CoP in brokering connections depended on how long they had spent in the community. Like Wilson et al. (2020) who identified that members who had greater seniority were less likely to engage the learning, more experienced members did not find the CoP valuable in supporting their connection to the wider science education community. This is perhaps reassuring as it shows time in the profession does allow for greater connectivity to the wider scientific community.

14.5.3 Research and Teaching Practice

Knowledge was created in the CoP by and between the members as they had authentic conversations online (Patton & Parker, 2017) about the theories, practices and policies of science education and their place within it. This was a combination of deliberate learning and learning that occurred organically through the interaction between the members. Some knowledge creation was intentional, such as having a purposely invited academic caution members to "narrow your focus, be strategic, think about what you commit to". At other times the learning was incidental as "we talked about our practice," identifying pre-service teaching challenges and negotiating curriculum reform. It was the "thinking together" that provided members with a means of "sustaining our practice" (Pyrko et al., 2017, p. 390). However, it was not easy to disentangle research and teaching practice, as we also researched within the domain of primary science education.

Ideas about practice were generated in the "chats" between each other but the extent to which they translated into practice depended upon ITE primary science teaching experience. For example, CoP members "new to primary science ... had lots of questions" and looked forward to getting "ideas" about practice from the online session, whilst a more experienced member initially rated the contribution of the CoP on their practice low but upon reflection agreed the online conversations did "inform practice" by contributing to an "iterative cycle" of practice and reflection. Overall, it was seen that talking together enabled us to build a shared vision about what good practice in science education in ITE was, and this subtly informed our teaching practice.

14.6 Conclusion and Recommendations

The online CoP assisted ECAs in extending their research and teaching practice capacities in primary science education. It enhanced their sense of belonging and identity as primary science education experts within New Zealand and for some, opened doors to the wider science community. The relationships between members enabled authentic conversations about their research and teaching practice and challenges in the ITE science education space. Positive impact on teaching practice was more marked for new appointees. The CoP had a higher effect on research capacities, which was not surprising due to research having higher status than teaching in academia (Mulholland et al., 2023; Sutherland et al., 2013).

The online CoP allowed isolated primary science education ECAs to discuss current trends and challenges in the primary science domain, and by extension sharpen their practice. This was important, as the lack of targeted professional development for academics in ITE has been identified as problematic (MacPhail et al., 2014). A recommendation to further enhance the efficacy of this CoP and others who may emulate it, is to include more discussion about teaching practice. This is especially important for less experienced ECAs. This could include talking about lesson design, content knowledge, pedagogical content knowledge, pertinent classroom activities, cultural perspectives, and literature to support learning.

It is well recognised that ITE teaching quality can affect pre-service teacher's confidence and ability to teach curricular subjects effectively (MacPhail et al., 2014). Therefore, it is necessary that engaging lessons be designed to support the development of pre-service teachers' confidence and ability to teach science competently after graduation (ECNZ, 2017; ERO, 2024). One way to enhance the quality and consistency of teaching experiences across different ITE providers is through forming a CoP with other primary science teacher educators with the aim of enhancing competency in the domain. This may meet the recommendation of ERO (2024) to "strengthen ITE programmes focusing on the areas where teachers are least prepared", which includes primary science (p. 121).

This study confirms that belonging to this CoP enabled members to deepen their research and teaching practice in the primary science education domain. Nevertheless, it must be acknowledged that these were the experiences of one small CoP and may not be transferable to other settings. However, this study aligns with other studies indicating CoPs are valuable in understanding social learning in educational settings and supporting members to learn and develop their practice (Adie et al., 2017; Camarao & Din, 2023). The findings contribute to the wider CoP field—specifically those exploring higher education, ITE, newly graduated teachers and science specialists. It is of particular interest to those who work in isolation or in small teams as it enlarges the possibility of obtaining domain targeted professional development through an online medium.

References

Adie, L., Mergler, A., Alford, J., Chandra, V., & Hepple, E. (2017). Teacher educators' critical reflection on becoming and belonging to a community of practice. In J. McDonald & A. Cater-Steel (Eds.), *Communities of practice: Facilitating social learning in higher education* (pp. 403–422). Springer.

Camarao, J., & Din, C. (2023). "A group of people to lean on and learn from": Graduate teaching assistant experiences in a pedagogy-focused community of practice [Article]. *Teaching and Learning Inquiry, 11*. https://doi.org/10.20343/teachlearninqu.11.11

Cohen, L., Manion, L., & Morrison, K. (2018). *Research methods in education* (8th ed.). Routledge. http://ebookcentral.proquest.com/lib/AUT/detail.action?docID=5103697

Cox, M. D. (2013). The impact of communities of practice in support of early-career academics [Article]. *International Journal for Academic Development, 18*(1), 18–30. https://doi.org/10.1080/1360144X.2011.599600

de Laat, M., Vrieling, E., & van den Beemt, A. (2017). Facilitation of social learning in teacher education: The 'dimensions of social learning framework'. In A. Cater-Steel & J. McDonald (Eds.), *Communities of practice: Facilitating social learning in higher education* (pp. 153–174). https://doi.org/10.1007/978-981-10-2879-3_7

Dickerson, C., White, E., Levy, R., & Mackintosh, J. (2021). Teacher leaders as teacher educators: Recognising the 'educator' dimension of some teacher leaders' practice. *Journal of Education for Teaching, 47*(3), 395–410. https://doi.org/10.1080/02607476.2021.1880250

Education Council New Zealand I Matatu Aotearoa. (2017). *Our code our standards*. https://teachingcouncil.nz/assets/Files/Code-and-Standards/Our-Code-Our-Standards-Nga-Tikanga-Matatika-Nga-Paerewa.pdf

Education Review Office Te Tari Arotake Matauranga. (2024). *Ready steady, teach: How prepared and supported are new teachers?* https://ero.govt.nz/our-research/ready-set-teach-how-prepared-and-supported-are-new-teachers

Hardy, A., Murrary, R., Thow, M., & Smith, M. (2022). 'So maybe I'm not such an imposter': Becoming an academic after a life as a teacher-practitioner. *Higher Education Research & Development, 41*(1), 163–176. https://doi.org/10.1080/07294360.2020.1835835

Hollywood, A., McCarthy, D., Spencely, C., & Winstone, N. (2020). 'Overwhelmed at first': The experience of career development in early career academics [Article]. *Journal of Further and Higher Education, 44*(7), 998–1012. https://doi.org/10.1080/0309877X.2019.1636213

Kensington-Miller, B. (2021). 'My attention shifted from the material I was teaching to student learning': The impact of a community of practice on teacher development for new international academics [Article]. *Professional Development in Education, 47*(5), 870–882. https://doi.org/10.1080/19415257.2019.1677746

Lunenberg, M., Dengerin, J., & Korthagen, F. (2014). *The professional teacher educator: Roles, behaviour, and professional development of teacher educators*. Sense Publishers.

MacPhail, A., Patton, K., Parker, M., & Tannehill, D. (2014). Leading by example: Teacher educators' professional learning through communities of practice. *Quest (00336297), 66*(1), 39–56. https://doi.org/10.1080/00336297.2013.826139

McDonald, J., & Cater-Steel, A. (2017). *Communities of practice: Facilitating social learning in higher education* (A. Cater-Steel & J. McDonald, Eds.). Springer. https://link.springer.com/book/10.1007/978-981-10-2879-

McDonald, J., Star, C., & Margetts, F. (2012). *Identifying, building and sustaining leadership capacity for communities of practice in higher education*. www.olt.gov.au

Mercieca, B. (2017). What is a community of practice. In J. McDonald & A. Cater-Steel (Eds.), *Communities of practice: Facilitating social learning in higher education* (pp. 3–26). Springer. https://doi.org/10.1007/978-981-10-2879-3_19

Mula, J., Rodríguez, C. L., Domingo Segovia, J., & Cruz-González, C. (2022). Early career researchers' identity: A qualitative review [Article]. *Higher Education Quarterly, 76*(4), 786–799. https://doi.org/10.1111/hequ.12348

Mulholland, K., Nichol, D., & Gillespie, A. (2023). 'It feels like you're going back to the beginning…': Addressing imposter feelings in early career academics through the creation of communities of practice [Article]. *Journal of Further and Higher Education, 47*(1), 89–104. https://doi.org/10.1080/0309877X.2022.2095896

Patton, K., & Parker, M. (2017). Teacher education communities of practice: More than a culture of collaboration [Article]. *Teaching and Teacher Education, 67*, 351–360. https://doi.org/10.1016/j.tate.2017.06.013

Pharo, E., Davison, A., McGregor, H., Warr, K., & Brown, P. (2014). Using communities of practice to enhance interdisciplinary teaching: Lessons from four Australian institutions. *Higher Education Research and Development, 33*(2). https://doi.org/10.1080/07294360.2013.832168

Price, E., Coffey, B., & Nethery, A. (2015). An early career academic network: What worked and what didn't. *Journal of Further and Higher Education, 39*(5), 680–698. https://doi.org/10.1080/0309877X.2014.971106

Pyrko, I., Dörfler, V., & Eden, C. (2017). Thinking together: What makes Communities of Practice work? [Article]. *Human Relations, 70*(4), 389–409. https://doi.org/10.1177/0018726716661040

Rienties, B., & Hosein, A. (2020). Complex transitions of early career academics (ECA): A mixed method study of with whom ECA develop and maintain new networks [Article]. *Frontiers in Education, 5*. https://doi.org/10.3389/feduc.2020.00137

Stewart, G., Eames, C., Hipkins, R., Cheng, M., Birdshall, S., Buntting, C., Carpendale, J., Edwards, R., Hunt, D., & Swanson, C. (2024). What's the future for science in the New Zealand curriculum? *New Zealand Journal of Educational Studies*. https://doi.org/10.1007/s40841-024-00345-3

Sutherland, K., Wilson, M., & Williams, P. (2013). *Success in academia? The experience of early career academics in New Zealand universities*. https://ako.ac.nz/knowledge-centre/academic-career-success/#:~:text=Key%20Findings&text=Research%20found%20there%20were%20too,with%20life%20and%20work%20experience

Swanson, C. J. (2021). Varied experiences of science education in Auckland primary schools: Siloed, integrated, or somewhere in between? *Teachers and Curriculum, 21*(2), 65–75. https://doi.org/10.15663/tandc.v21i0.365

Teaching Council New Zealand I Matatu Aotearoa. (2019). *ITE Programme approval, monitoring and review requirements.* https://teachingcouncil.nz/assets/Files/ITE/ITE-Programme-Approval-Monitoring-and-Review-Requirements.pdf

Teaching Council of Aotearoa New Zealand. (2022). *Creating an equitable future-focused Initial Teacher Education (ITE) system: The new ITE requirements—The journey so far*. https://teachingcouncil.nz/assets/Files/Publications/TC-ITE-Journey-publication-22-Feb-2022.pdf

Teaching Council of Aotearoa New Zealand. (2023). *List of approved ITE providers.* https://teachingcouncil.nz/professional-practice/ite-providers/approved-ite-programmes/

van Lankveld, T., Schoonenboom, J., Beishuizen, J., Croiset, G., & Volman, M. (2017). Developing a teacher identity in the university context: A systematic review of the literature [Review]. *Higher Education Research and Development, 36*(2), 325–342. https://doi.org/10.1080/07294360.2016.1208154

Weinberg, A. E., Balgopal, M. M., & Sample McMeeking, L. B. (2021). Professional growth and identity development of stem teacher educators in a community of practice. *International Journal of Science and Mathematics Education, 19*, 99–120. https://doi.org/10.1007/s10763-020-10148-9

Wenger, E. (1998). *Communities of practice: Learning, meaning, and identity.* Cambridge University Press.

Wenger, E. (2018). A social theory of learning. In K. Illeris (Ed.), *Contemporary theories of learning: Learning theorists … in their own words* (2nd ed., pp. 209–218). Routledge. https://doi.org/10.4324/9781315147277

Wenger, E., McDermott, R., & Snyder, W. (2002). *Cultivating communities of practice: A guide to managing knowledge.* Harvard Business School Press.

Wenger-Trayner, E., & Wenger-Trayner, B. (2015). *An introduction to communities of practice: A brief overview of the concept and its uses.* http://wenger-trayner.com/introduction-to-communities-of-practice/

Wilson, A., Wilson, C., & Witthaus, G. (2020). Using a community of practice in higher education: Understanding the demographics of participation and impact on teaching. *International Journal of Teaching and Learning in Higher Education, 32*(1), 39–48. https://www.isetl.org/ijtlhe/pdf/IJTLHE3644.pdf

Chapter 15
Conclusion: Stories of Possibility and Access for Primary Science Education

Sarah Earle, Helen Georgiou, Christine Preston, and Ange Fitzgerald

Abstract This concluding chapter brings together key themes from across the volume. To counter a deficit narrative, we draw on stories of practice from across the preceding chapters to present possibilities and opportunities for primary science teaching. An emergent overarching theme is the exploration of accessibility and inclusivity in learning in primary science from the perspectives of children, teachers and communities. The importance of multimodality, teacher professional learning and the development of STEM teacher identity were also identified as key conceptual threads running through this collection. The context-dependent nature of primary science is exemplified in discussions of science teaching in diverse and under-represented communities. The mosaic of perspectives is brought together as an assemblage, providing an evidence-informed basis for decision making in the sector

Keywords Primary science · Multimodality · Teacher identity · Access · Assemblage

15.1 Introduction

Readers of this collection have engaged with a rich array of conceptual understanding, research methodologies and evidence-informed outcomes, all linked by a common thread of improving access and enriching understandings of science

S. Earle (✉)
Bath Spa University, Bath, UK
e-mail: s.earle@bathspa.ac.uk

H. Georgiou
University of Wollongong, Wollongong, NSW, Australia

C. Preston
University of Sydney, Sydney, NSW, Australia

A. Fitzgerald
RMIT University, Melbourne, Australia

S. Earle et al. (eds.), *Primary Science Learning for Children, Teachers, and Communities*, https://doi.org/10.1007/978-981-95-0561-6_15

concepts to inform primary science learning and teaching. Through the stories of educators and education researchers who have successfully navigated challenges, this collection offers practical strategies and perspectives that can be adapted to various educational settings. We are particularly passionate that this edited collection speaks back into the widespread deficit discourse that often dominates and ultimately frames primary science education as a field fraught with challenges and shortcomings (Fitzgerald & Smith, 2016). We were also keen to provide a platform for underrepresented voices, particularly those from non-metropolitan and marginalised communities. By focusing on the successes within the field, we hope to empower educators to think creatively and reimagine the potential of their own practice.

Supporting children, teachers and communities to access primary science has emerged as an overarching theme in this volume. Primary school teaching is complex, with teachers navigating a wide array of subjects, including some where they feel less confident, this being science in many cases. Thus, considerations of access to science are not solely student-focused, teacher professional development and community perspectives of science are also at play. The stories of practice in this text provide context-rich explanations from the point of view of children, teachers and community, and in this concluding chapter, we revisit these perspectives, to consider how access to science or Science, Technology, Engineering and Mathematics (STEM) can be facilitated in each context. The emergent key threads of multimodality, teacher identity and inclusivity will be foregrounded as we draw on examples from the chapters.

15.2 Science or STEM?

The boundaries of primary science intentionally remained flexible within this book, to reflect the context-dependent definitions of what would be contained within the subject. As noted in Chapter 1, in some education systems, primary science is focused on the traditional content of biology, chemistry and physics, whilst others explicitly acknowledge the relationships between Science, Technology, Engineering and Mathematics (STEM). The benefits of further cross-curricular links can also be considered, for example, with topic or project-based learning, science to support language development, or linking science with the Arts in a STEAM approach. Whilst fruitful cross-curricular links are encouraged, and stories of practice rarely sit neatly in subject silos, there is also an argument for a continuing focus on science within an interdisciplinary context, so that its integrity remains (Tytler, 2020). Some also argue that effective interdisciplinary learning must be grounded in solid disciplinary knowledge (AERO, 2024). Without a shared definition for STEM (Akerson & Bartels, 2023), educational practices may often reasonably, flexibly and usefully be considered both individually and through an interdisciplinary lens, as in this book.

The interplay between science and STEM (or STEAM) was a recurring theme across many chapters, with practitioners balancing interdisciplinary learning with a particular focus on science. The embedded nature of the definition of science on occasion led to its conflation during the writing process. For example, a design and technology example proposed to exemplify teaching in 'science' was a natural addition for an author in a STEM context, but less so for an author in a context with narrower subject silos. Explicit discussion about what is included under the science umbrella is necessary to avoid miscommunication and perhaps to understand why STEM is a more natural grouping in many contexts. Educators may also argue for integration of the Arts to become STEAM, to increase creative processes and emphasize humanistic dimensions (Gavrilas & Kotsis, 2025).

We see opportunities for learning across a range of international contexts within this discussion. What is included in primary science teaching can stretch from the more narrow subject areas of biology, chemistry and physics, to the broader STEM groupings of natural sciences, that include geography and geology, and even the social sciences in countries that take an integrated subject view such as Switzerland. Comparisons across countries, within and between chapters, can provide those with narrower definitions, opportunities to explore cross-curricular stories of practice; whilst those with broader enactments can utilise stories of science lessons to explore the focused teaching of specific scientific concepts. Of course in primary science, we often have the same teacher working with the class for multiple subjects, making disciplinary boundaries less conspicuous. The cross-curricular nature of primary school teaching supports real-world applications, with the child exploring the world around them free of subject boundaries. For example, biomedical problem solving was described in Chapter 4's stories of practice. Nevertheless, we also want to support children to appreciate science as a discipline, for them to see science as 'for them' (Nag Chowdhuri et al., 2021), something that they can relate to and be part of as a global citizen now and in the future.

15.3 Stories and Possibilities for Children

A cross-cutting theme within the children-focused section of this book is improving access to primary science through multimodality (Kress, 2010), which concerns meaning-making and communication across a variety of modes such as speech, gesture, images, writing and objects. In Chapter 4, Klapwijk and Zhu demonstrate the importance of both manipulating practical materials and their representations to support spatial reasoning. Likewise, Meidell Sigsgaard considers the interplay between hands-on activities and language learning in Chapter 5, whilst Yeo and Tan discuss the relation between representational skills and the learning of scientific concepts when utilizing images in Chapter 3. In all of these chapters, it is noted that the relationship between different modalities is not simple, it cannot be assumed that a hands-on activity or presentation of an image will lead to the development of scientific understanding. '*Explanatory ideas do not 'emerge' from observations... ideas have to be introduced*' (Abrahams & Millar, 2008, p. 1965). Connecting the ideas, linking practical or visual aids to the scientific concepts, may require teacher intervention, such as in the *image-to-writing* (I2W) approach (Yeo et al., 2021), which moves from exploring a phenomenon to the creation of images and writing. This sense of moving from one mode to another, from concrete to more abstract, is also visible when considering language development in Chapter 5, moving from doing hands-on science activities, to understanding the scientific concepts via scaffolds to develop language use. The transformative and transductive processes of moving from one modality to another are identified as helpful to learning. For example in Chapter 4, when creating a food chains diagram, identifying the relationship between the organisms as they are placed in the food web supports the development of the concept of interdependence. Making spaces for choices in modality, or the timing in moving from one mode to another, provides an opportunity for children's agency in primary science, as they select the means for interacting and sharing their science ideas.

In primary science, consideration of multimodality is particularly pertinent, as we identify the purposeful use of practical work (Earle et al., 2025). Science for young children begins with the natural world, exploring materials, objects and phenomena to make sense of the way that the world works and to develop the language to describe it. The *show, slow* and *spontaneous* science opportunities presented by Petersen, Ahrenkiel and Jørgensen in Chapter 2 arose from consideration of practice in early childhood settings, but such opportunities are also recognisable when working with older children. The *slow* science of the frog life cycle was facilitated by practitioners in the early childhood setting, with continued opportunities to observe, discuss and communicate. Such *slow* science is also exemplified in Chapter 6, with Earle and Grob describing a 'chick diary', whereby observations of a growing young chicken are recorded in writing and drawing. Teacher discussion and feedback prompts are utilized in this case to further develop observation skills. Such formative use of assessment is also seen in stories of practice in other chapters. For example, Chapter 3 presents a sequence of 'explore, work-on, review' to support understanding of the concept of temperature. Involving children explicitly within the reviewing provides another opportunity for agentic development, with children able to provide feedback to peers as they inquire together. The cyclical nature of formative assessment, presented in Fig. 6.1, is also implicit in the iterative cycles of research presented in multiple chapters.

The prevalence of design-based research (Barab & Squire, 2004; Anderson & Shattuck, 2012), is evident across many chapters within the Children section of this book. In recognition of the complexity of primary science classrooms, together with raising the profile of participants as agentic co-investigators, design-based research iteratively co-constructs theory and products in developmental cycles to test interventions within real contexts. For example in Chapter 5, applying the concept of semantic gravity from Legitimation Code Theory (Maton, 2014), researchers and teachers were able to tailor activities to initiate a move from language with stronger semantic gravity (context-dependent, everyday descriptions) to weaker semantic gravity (more generalisable, scientific explanations). Design research is also visible in Chapter 3's cycles to develop the concept of temperature, Chapter 4's testing of strategies to support spatial reasoning and Chapter 6's iterative development of formative assessment. In each case, design-based research is utilised to address an ongoing issue in science education, such as bridging the gap between hands-on activities and conceptual development (Abrahams & Millar, 2008) by integrating language development in Chapter 5, or addressing the difficulty of implementing formative assessment (Furtak et al., 2016) by exemplification in Chapter 6.

15.4 Stories of STEM Teacher Identity and Inclusive Practice

Two underlying themes within the Teacher section of this book are the formulation of an integrated STEM teacher identity and consideration of a holistic approach to inclusive science teaching. STEM teacher identity describes how primary teachers enrich their persona from simultaneously *learning* more STEM subject knowledge and developing practices for *teaching* Science, Technology, Engineering and Mathematics. Critically the term integrated here does not refer to combining discipline areas, it emphasises the dual roles as a STEM learner and teacher (Galanti & Holincheck, 2022). Primary teachers are usually seen (and see themselves) as generalist practitioners. Encapsulating an integrated STEM teacher identity within their primary teacher identity is a complex transformation that requires support through professional development. Primary teachers also bring an inclusive and holistic lens enabling them to see links with additional learning areas and bring them into STEM which can result in theme-based pedagogical approaches. Each chapter in the Teacher section tells a story of primary teachers developing an identity beyond being a generalist teacher in which they develop professional competencies and personal dispositions towards STEM (Falloon et al., 2020; Fitzpatrick, 2024; Leavy et al., 2023) as inclusive practice. The development of such identities builds a new sense of agency, where teachers feel empowered to make actionable decisions in their pedagogical choices (Biesta et al., 2015).

As highlighted throughout this volume, primary teachers often report feeling hesitant to teach science. In Chapter 7, Nash, Bradbury and Fitzgerald describe the transformative experiences from an immersive STEM course during initial teacher training in which primary pre-service teachers reflected that STEM was not on their radar. Chapter 8 by Keijzer, Embrechts, Hotze, Jonker, Bakker, Boels and van der Zee outlines a national initiative to address the needs of STEM education in the Netherlands. Chapter 8 describes Professional Learning Networks of teachers and researchers, enabling in-service teachers to meet the challenges of assessing learning outcomes, relating mathematics with other STEM domains, data literacy and maker education. In Chapters 9 and 10, the lack of professional development opportunities for primary teachers of science is highlighted. In Chapter 9, Way, Cartwright, Preston and Galstaun demonstrate how an evidence-based professional learning program can inspire Australian primary teachers to engage their students in diverse, innovative interdisciplinary STEM projects, unique to their local contexts. Through Chapter 10, Crook and Georgiou relate case studies of STEM professional development with primary teachers in school contexts to meet situational factors, including teacher, student and systemic needs. The chapters allude to challenges in STEM education in elementary classrooms and how novel professional development approaches with both pre-service and practising teachers can lead to positive outcomes and increased access.

Collectively, the chapters show pre-service and practising primary teachers engaged with STEM, through a holistic lens. Stinken-Rösner et al. (2020) highlight that inclusivity in science education constitutes pedagogies which 'acknowledge diversity, recognise barriers, and enable participation' by all (p. 35). They propose a scheme of inclusive science teaching and learning that includes 'reasoning about scientific issues, learning science content, doing science, and learning about science' (p. 35). Across all these four chapters, the inclusivity of STEM and inquiry-based pedagogies is emphasised. In Chapter 7, the power of STEM as a holistic framework which encourages authenticity, facilitates engagement with real world problems and seamlessly fits into a necessarily broad primary curriculum is highlighted. In this chapter, the framework of the Theory of Practice Architectures showcases the different layers of interactions that impact assessing STEM and inquiry skills, emphasising the need to have a way of measuring more holistically, the outcomes of science education. Chapter 7 offered insights into how the course design and interactions within the learning environment influenced pre-service teacher outcomes in Australia. By addressing cultural, material and social factors, a university course can challenge existing narratives, offer practical experiences and cultivate a positive attitude towards STEM teaching. Carefully considering these factors could help in equipping pre-service generalist teachers with the knowledge, skills and attitudes needed to integrate STEM into their future classrooms (Kartal & Tasdemir, 2021).

In both Chapters 8 and 9, authors draw on models of effective professional development to design and reflect on extensive programs for primary STEM teachers. In Chapter 8, the STEM Academy, which reached 500 individual teachers from 147 schools, presents a 'tri-level' approach, beginning with short, accessible STEM 'experiences', building up to iteratively designed real-world projects that are implemented by students. The authors report that this levelled approach empowered teachers to engage other teachers and students in STEM in their schools. STEM projects that involve real problems, real solutions and local contexts are showcased, for example, the Grandbuddy project in Chapter 9 and Sensory Garden design in Chapter 10. In these chapters, STEM projects empowered progressive development of skills and content knowledge for teachers, and inclusively engaged primary students.

15.5 Stories of Communities

Moving into the final chapters of this book signals a shift further outwards. Not simply into considerations of communities, such as those local communities described by Zulu, Nkambale, Sibanda and Fitzgerald in Chapter 11, and the Community of Practice explored by Swanson and Saunders in Chapter 14; but also into 'alternative spaces', as McCann, White and Dionne in Chapter 12 and Cirkony, Kim, Tweed, Carpendale and Preston in Chapter 13 respectively unpacked. These spaces offer a way to challenge the hegemony of science education by asking readers to consider different ideas, structures and political and cultural perspectives in order to facilitate access to science for otherwise marginalised groups.

Discourses of science, whether they are textbook representations of an unchangeable, detached and objective science (McComas et al., 2020), or curricula that focus on white male science (Ross et al., 2023), have contributed to a reputation of elitism and exclusivity. Romanticised and over-simplified descriptions of science similarly fail to capture its social and cultural influences, its humanity, and its fallibility (Georgiou, 2023). In an effort to challenge these assumptions and reimagine science education as more accessible, in this volume, a range of approaches, such as story-based approaches (Chapter 12), permaculture practice (Chapter 12), culturally-responsive teaching (Chapter 13) or place-based approaches (Chapter 12) to science education are presented. In particular, the weaving of Western Science and Indigenous knowledge systems are exemplified, to challenge existing power dynamics, and show how value can be shared amongst different epistemologies within science. One simple but powerful example in Chapter 13, outlines how the trophic pyramid, with apex predators at its peak, use ascending levels to show both power, in that apex predator at the top of the food chain, and vulnerability, in that apex predators rely on so many other organisms for their survival. This bilateral conceptualisation both acknowledges and challenges existing narratives about the natural world. These chapters therefore not only offer a range of diverse approaches to teaching (e.g., multigrade classrooms in Chapter 11) and research (e.g., narrative Inquiry in Chapter 13), but also a broader call to action, as empowered agents of change, to question and challenge the status quo.

15.6 Implications and Future Directions

In concluding this collection, the primary science education 'stories' underscore the intricate, multifaceted journey that teachers, teacher educators, policymakers and communities must undertake to foster scientific literacy and inquiry in young learners. The chapters reveal a mosaic of perspectives, each adding to a complex yet coherent tapestry of what science education can and should achieve in primary settings. This concluding chapter has highlighted the implications for diverse audiences and charts a path for practice and research that embraces complexity and context, emphasising that effective primary science teaching cannot be confined to siloed subject-focused disciplines but must integrate knowledge from various sources.

For teachers, the collection advocates for a broadened identity, encouraging them to see themselves not merely as science instructors but as facilitators of an interdisciplinary, child-centered approach. This collection suggests a shift in professional identity towards becoming facilitators of science and STEM inquiry and problem-solving, cultivating curiosity and fostering skills that transcend scientific knowledge alone. Teacher identity formation, as illustrated in several chapters focused on professional learning and identity development, demonstrates how in-service and pre-service teachers can grow more confident in delivering integrated STEM lessons that are relevant and engaging. For practice, the recommendation is clear: teachers need continued, context-specific professional learning that honours their experiences and addresses classroom realities, enabling them to embed science within a wider educational narrative.

Teacher educators are called upon to design learning experiences that incorporate inclusive pedagogies. These must be responsive to the realities of classrooms, where learning is often chaotic, non-linear, and highly contextualized. An assemblage approach (Strom, 2015), as employed across chapters in this collection as a sense making device, acknowledges that primary science education is informed by an array of theories, from multimodality to design-based research. No single framework can capture the full range of experiences in primary science education, nor should it. Instead, the notion of 'assemblage' reveals the richness of drawing upon multiple frameworks, allowing educators to be adaptive and responsive. The recommendation here is to design educator programs that introduce both pre-service and in-service teachers to a variety of approaches, enhancing their ability to synthesize these methods and adapt to their students' needs.

For policymakers, this work advocates for policy that acknowledges the inherently interdisciplinary nature of primary science. Rigid curriculum structures may hinder teachers' ability to adapt to classroom dynamics, limiting their capacity to explore real-world applications that blur disciplinary lines. Recognising the importance of flexibility within curriculum guidelines can empower teachers to integrate science seamlessly with other subjects, supporting holistic learning that reflects the real-world interconnectedness of scientific knowledge. A recommendation for research policy is to support studies that focus on diverse, cross-disciplinary approaches in science education, with special attention to inclusive, context-based and child-centred learning.

As a whole, this collection acknowledges the inherent 'messiness' of primary science education. It is a field where clear-cut definitions and prescribed outcomes can often overlook the nuanced and unpredictable nature of learning across the primary years. Embracing this complexity rather than seeking to resolve it is critical. Recognising that effective primary science involves a blend of practical engagement, representational skills, language development and interdisciplinary inquiry supports an education that is grounded in the natural curiosity of children. Such an approach sees each learning experience as an opportunity for students to connect with science in meaningful ways.

As we consider the future of primary science education, it is essential to acknowledge the Anthropocene as a driving context for how and why we teach science. The environmental and socio-scientific challenges facing today's learners demand educational responses that are rooted in systems thinking, ethics, and hope. Primary science education has a crucial role to play in nurturing these capacities from an early age.

Stories of practice across this collection demonstrate how local, place-responsive, and interdisciplinary approaches can empower children, teachers and communities to meaningfully engage with these global issues. Integrating this awareness into curricula and pedagogy ensures that science learning and teaching is not only relevant but transformative by helping students connect knowledge to action and imagine new possibilities for more sustainable futures (Wallace et al., 2022).

Finally, the future of primary science education lies in continuing to explore this breadth of methods, embracing ambiguity and viewing challenges as opportunities for innovation. The collection reaffirms the power of storytelling as a pedagogical tool, emphasising the value of participatory and strengths-based approaches. By weaving together these learning stories from children, teachers and communities, this work invites ongoing inquiry and adaptive practice, ensuring that primary science education remains vibrant, inclusive and responsive to the changing needs of society.

References

Abrahams, I., & Millar, R. (2008). Does practical work really work? A study of the effectiveness of practical work as a teaching and learning method in school science. *International Journal of Science Education, 30*(14), 1945–69. https://doi.org/10.1080/09500690701749305

Akerson, V. L., & Bartels, S. L. (2023). Elementary science teaching: Toward the goal of scientific literacy. In *Handbook of research on science education* (pp. 528–558). Routledge. https://doi.org/10.4324/9780367855758

Anderson, T., & Shattuck, J. (2012). Design-based research: A decade of progress in education research? *Educational Researcher, 41*(1), 16–25.

Australian Education Research Organisation (AERO). (2024). *A knowledge-rich approach to curriculum design.* https://www.edresearch.edu.au/research/research-reports/knowledge-rich-approach-curriculum-design

Barab, S., & Squire, K. (2004). Design-based research: Putting a stake in the ground. *The Journal of the Learning Sciences, 13*(1), 1–14. https://doi.org/10.1207/s15327809jls1301_1

Biesta, G., Priestley, M., & Robinson, S. (2015). The role of beliefs in teacher agency. *Teachers and Teaching, 21*(6), 624–640. https://doi.org/10.1080/13540602.2015.1044325

Earle, S., Read, S., Bianchi, L., & Jordan, J. (2025). *Purposeful practical work in primary science.* Bath Spa University & The University of Manchester. https://www.bathspa.ac.uk/projects/practical-work-in-primary-school-science/

Falloon, G., Hatzigianni, M., Bower, M., Forbes, A., & Stevenson, M. (2020). Understanding K-12 STEM education: A framework for developing STEM literacy. *Journal of Science Education and Technology, 29*, 369–385. https://doi.org/10.1007/s10956-020-09823-x

Fitzgerald, A., & Smith, K. (2016). Science that matters: Exploring science learning and teaching in primary schools. *Australian Journal of Teacher Education, 41*(4), 64–78.

Fitzpatrick, M. (2024). Drawing the past to envision the future: Supporting the development of primary STEM teacher identity. *Irish Educational Studies*, 1–24. https://doi.org/10.1080/03323315.2024.2359693

Furtak, E. M., Keimer, K., Circi, R. K., Swanson, R., deLeon, V., Morrison, D., & Heredia, S. (2016). Teachers' formative assessment abilities and their relationship to student learning: Findings from a four-year intervention study. *Instructional Science, 44*, 267–291.

Galanti, T. M., & Holincheck, N. (2022). Beyond Content and Curriculum in Elementary Classrooms: Conceptualizing the Cultivation of Integrated STEM Teacher Identity. *International Journal of STEM Education, 9*(1), 1–10. https://doi.org/10.1186/s40594-022-00358-8

Gavrilas, L., & Kotsis, K. T. (2025). The evolution of STEM education and the transition to STEAM/ STREAM. *Aquademia, 9*(1), ep25002. https://doi.org/10.29333/aquademia/16313

Georgiou, H. (2023). Preservice teachers' views of the "social embeddedness" tenet of the nature of science: A new method of analysis. *Journal of Science Teacher Education, 34*(3), 248–266.

Kartal, B., & Tasdemir, A. (2021). Pre-service teachers' attitudes towards STEM: Differences based on multiple variables and the relationship with academic achievement. *International Journal of Technology in Education (IJTE), 4*(2), 200–228. https://doi.org/10.46328/ijte.58

Kress, G. (2010). *Multimodality: A social semiotic approach to contemporary communication.* Routledge.

Leavy, A., Dick, L., Meletiou-Mavrotheris, M., Paparistodemou, E., & Stylianou, E. (2023). The prevalence and use of emerging technologies in STEAM education: A systematic review of the literature. *Journal of Computer Assisted Learning, 39*(4), 1061–1082. https://doi.org/10.1111/jcal.12806

Maton, K. (2014). *Knowledge and knowers: Towards a realist sociology of education.* Routledge.

McComas, W. F., Clough, M. P., & Nouri, N. (2020). Nature of science and classroom practice: A review of the literature with implications for effective NOS instruction. In: McComas, W. F. (eds) *Nature of Science in Science Instruction. Science: Philosophy, History and Education.* Springer, Cham. https://doi.org/10.1007/978-3-030-57239-6_4

Nag Chowdhuri, M., King, H., & Archer, L. (2021). *The primary science capital teaching approach: Teacher handbook.* UCL Institute of Education.

Ross, K., Galaudage, S., Clark, T., Lowson, N., Battisti, A., Adam, H., Ross, A. K., & Sweaney, N. (2023). Invisible women: Gender representation in high school science courses across Australia. *Australian Journal of Education, 67*, 231–252.

Stinken-Rösner, L., Rott, L., Hundertmark, S., Baumann, T., Menthe, J., Hoffmann, T., Nehring, A., & Abels, S. (2020). Thinking inclusive science education from two perspectives: Inclusive pedagogy and science education. *RISTAL, 3*, 30–45.

Strom, K. (2015). Teaching as assemblage: Negotiating learning and practice in the first year of teaching. *Journal of Teacher Education, 66*(4), 321–333. https://doi.org/10.1177/0022487115589990

Tytler, R. (2020). STEM education for the twenty-first century. In *Integrated approaches to STEM education: An international perspective* (pp. 21–43). Springer.

Wallace, M. F. G., Bazzel, J., Higgins, M., & Tolbert, S. (2022). *Reimagining science education in the Anthropocene.* Palgrave Macmillan.

Yeo, J., Lim, E., Tan, K. C. D., & Ong, Y. S. (2021). The efficacy of an image-to-writing approach to learning abstract scientific concepts: Temperature and heat. *International Journal of Science and Mathematics Education, 19*, 21–44. https://doi.org/10.1007/s10763-019-10026-z

GPSR Compliance

The European Union's (EU) General Product Safety Regulation (GPSR) is a set of rules that requires consumer products to be safe and our obligations to ensure this.

If you have any concerns about our products, you can contact us on ProductSafety@springernature.com

In case Publisher is established outside the EU, the EU authorized representative is:

Springer Nature Customer Service Center GmbH
Europaplatz 3
69115 Heidelberg, Germany

Batch number: 09097872

Printed by Printforce, the Netherlands